The History of the Geometry Curriculum in the United States

a volume in
Research in Mathematics Education

Series Editor:
Barbara J. Dougherty
University of Mississippi

Research in Mathematics Education

Barbara J. Dougherty, Series Editor

The History of the Geometry Curriculum in the United States (2008)
by Nathalie Sinclair

The Classification of Quadrilaterals: A Study in Definition (2008)
edited by Zalman Usiskin

The Intended Mathematics Curriculum as Represented in State-Level Curriculum Standards: Consensus or Confusion? (2006)
edited by Barbara Reys

The History of the Geometry Curriculum in the United States

by

Nathalie Sinclair
Simon Fraser University

Information Age Publishing, Inc.
Charlotte, North Carolina • www.infoagepub.com

Library of Congress Cataloging-in-Publication Data

Sinclair, Nathalie.
The history of the geometry curriculum in the United States / by Nathalie Sinclair.
p. cm. — (Research in mathematics education)
Includes bibliographical references.
ISBN-13: 978-1-59311-696-5 (paperback)
ISBN-13: 978-1-59311-697-2 (hardcover)
1. Geometry--Study and teaching—History—United States.
2. Geometry—History—United States. I. Title.
QA461.S56 2008
516.0071'073—dc22

2007047730

ISBN 13: 978-1-59311-696-5 (paperback)
ISBN 13: 978-1-59311-697-2 (hardcover)

Printed in the United States of America

CONTENTS

Acknowledgments *vii*

1. Introduction *1*

2. The "Big Bang" of Geometry: Euclid's *Elements* *5*

The Implicit Aims of Euclid's Geometry: Why? *7*

The Scope and Sequence of Euclid's Geometry: What? *8*

The Methods and Instruments of Euclid's Geometry: How? *9*

3. Steps in the History of the Geometry Curriculum *13*

The Geometry Curriculum in the United States in the Early Nineteenth Century *14*

1844: Geometry Lands on the List of University Entrance Requirements *18*

1850s: Early Pedagogical Influences on the Geometry Curriculum *19*

1860s: A Mathematical Challenge to the Dominance of Euclidean Geometry *23*

1892: First Attempts to Standardize the School Geometry Curriculum *26*

1902: Perry and Moore and Theories of Learning *33*

1920s and 1930s: Committees and Their Reports Redux *37*

1935: The Bourbaki Group and its Impact on Geometry *46*

1944: Post-War Developments *53*

1955: Beginning of the "Era of Reform" *56*

The 1960s: Time for Transformations *63*

1975: The NACOME Report Redefines "Basic Skills" *68*

The 1980s: The De-Degradation of "Geometric Consciousness" *72*
1989 NCTM: *Curriculum and Evaluation Standards for School Mathematics* *78*
1991: Dynamic Geometry Software *82*
2000: New and Improved Principles and Standards *86*

4. Where Are We Now and Where Are We Going? ***91***

References ***95***

About the Author ***105***

ACKNOWLEDGMENTS

I would like to thank Patrick Callaghan for access to his impressive collection of historical geometry textbooks. I would also like to thank Zalman Usiskin for so generously sharing his extensive insights and experience. Finally, I thank my reviewers—Tom Banchoff, Doris Schattschneider, Zalman Usiskin, Glenda Lappan, and Chris Hirsch—for their careful attention and thoughtful responses.

CHAPTER 1

INTRODUCTION

This monograph has the luxury of investigating the evolution of the geometry curriculum in the United States over the past 150 years, a luxury that no regular-length article can afford. Other scholars who have written histories of mathematics curricula have impressed upon their readers the benefits that such endeavors can have, such as preventing mistakes made repeatedly in the past by curriculum developers, policymakers, teachers, and educational researchers. My goal is somewhat different, and perhaps less ambitious, in that I hope to increase the mathematics education community's awareness of the shape and nature of the geometry curriculum as it currently stands by investigating both its historical roots and the many events and influences that have led to its present state. By revisiting past decisions and debates, I aim to encourage readers to think about how things might have turned out differently—Could grade 10 geometry have been a course devoted to solid geometry? Could the algebra course have shouldered the responsibility for teaching proof?—and how things may yet change in the coming decades.

Over the past century, curriculum scholars have dedicated books, book chapters, and journal articles to the history of geometry teaching and learning. Some have taken an international scope, whereas others have focused more narrowly on developments in specific countries. Surely, many of these scholars would have wished for time travel opportunities during which they could actually sit in a 10th-grade geometry class in Michigan at the turn of the 20th century as the teacher introduced

The History of the Geometry Curriculum in the United States
pp. 1–3

Euclid's postulates, or in a junior high "intuitive geometry" class in Texas in the 1950s as students measured the angles of a triangle. But such first-hand accounts of the geometry curriculum in action—the implemented curriculum—throughout the century can only tempt us. Instead, in order to inquire into the ways that geometry teaching and learning have changed in the past century, we must rely on more durable materials. Textbooks, especially in the United States, provide a good record of the intended curriculum at any given point in time (although one must keep in mind that the intended or implemented curriculum can vary significantly from the ideal or achieved curriculum). Policy documents can provide insight into the choices made in textbooks by hinting at the issues considered and the recommendations made by leaders in the field. Of course, since policy documents do not always lead to changes in textbooks, they can have little effect on what actually happens in those classrooms in Michigan or Texas. But other accounts can be brought to bear: journal articles in *The Mathematics Teacher*, for example, have provided firsthand accounts of teaching experiences, opinions, and arguments from those in the mathematics community, as well as reports on research studies related to the teaching and learning of geometry.

Once all these materials are gathered and digested, how does one synthesize them? A common strategy involves providing overviews by equally sized chunks of time. Thus in Donogue (2003), we find the following chapter headings: 1900–1920, 1921–1940, and so on. One problem with starting this kind of inquiry at the turn of the 20th century is that so many of the issues at stake—whether they are implicit or explicit—have their roots in much older decisions and events. And even if they did not, notable change (and stagnation) in mathematics education does not necessarily follow the neat numerical divides of the decades. Stamper (1909) combined the chunk-of-time strategy with a geographical one, analyzing developments in various countries (such as France, England, and the United States) separately. This provides a helpful national comparison, sometimes revealing some of the taken-for-granted assumptions and traditions found in our own countries, and other times emphasizing truly thorny issues that confront all involved in mathematics education. Yet the narrative thread becomes harder to follow as local conditions are studied in depth, without necessarily shedding light on more global developments. One could proceed by theme, and trace how, say, the teaching of plane and solid geometry was split or fused over the decades, or how the number of postulates and propositions increased and decreased over time. This might help contemporary curriculum developers become familiar with the issues involved in specific decisions, but it can also obscure the important ways in which changes involving one theme are interrelated with changes involving another, and indeed, with changes that may seem tangential to the issues at hand.

While no single approach is best, I have chosen to organize this monograph according to a chronological sequence of "notable events," as I see them, events that led to discernible changes in thinking about the geometry curriculum over the past century and a half—roughly the extent of time during which geometry has been taught in American schools. Notable events include important reports or commissions, influential texts, new schools of thought, and developments in learning technologies. Such notable events affected, among other things, the content and aims of the geometry curriculum; the nature of mathematical activity as construed by both mathematicians and those involved in mathematics education; and the kind of resources given to students for engaging in mathematical activity. Before embarking upon this train of notable events, it is worth taking a look at the "big bang" of geometry, namely, the moment in time that shaped—to a large degree—the future life of the geometry curriculum. This corresponds to the emergence of Euclidean geometry, the ancient Greek geometry developed over two millennia ago. Given its influence on the shape of the geometry curriculum, I propose that becoming familiar with the nature of the geometry articulated in Euclid's *Elements* can foster deeper understanding of the many tensions that have since surrounded the school geometry curriculum.

CHAPTER 2

THE "BIG BANG" OF GEOMETRY

Euclid's *Elements*

Euclid's *Elements* consists of 13 books written by the mathematician Euclid around 300 BCE. Although most of the theorems and constructions included in it had been developed earlier by other Greek mathematicians, Euclid's great achievement was to provide a systematic development in which a small set of axioms were used as a basis for deducing increasingly sophisticated results in plane and solid geometry. Throughout most of its 2,300-year history, Euclid's *Elements* was the principal geometry text studied by students in Europe. Between 1482 and 1900, there were more than 1,000 editions printed. In his preface to the first English translation, Billingsley (1570) wrote, "Without the diligent studie of Euclides Elements, it is impossible to attaine unto the perfecte knowledge of Geometrie, and consequently of any of the other Mathematical Sciences" (cited in Hartshorne, 2000, p. 460). In the preface to his 1798 edition, Bonnycastle said:

> Of all the works of antiquity which have been transmitted to the present time, none are more universally and deservedly esteemed than the Elements of Geometry which go under the name of Euclid. In many other

The History of the Geometry Curriculum in the United States
pp. 5–12

> branches of science the moderns have far surpassed their masters; but, after a lapse of more than two thousand years, this performance still retains its original preeminence, and has even acquired additional celebrity for the fruitless attempts which have been made to establish a different system. (1798, cited in Hartshorne, 2000, p. 460)

And Heath, in the preface to his definitive 1926 English translation, announced that:

> Euclid's work will live long after all the text-books of the present day are superseded and forgotten. It is one of the noblest monuments of antiquity; no mathematician worthy of the name can afford not to know Euclid, the real Euclid as distinct from any revised or rewritten versions which will serve for schoolboys or engineers. (p. vii)

The first book of the *Elements* begins with 23 definitions for geometric objects such as point, line, and surface, followed by five postulates, five axioms, and then a series of propositions logically deduced from the postulates and axioms (also known as common notions). Each subsequent book also begins with its own list of definitions, but the propositions can logically be deduced from previous books. The definitions, postulates, axioms, and propositions are stated without comment and ordered according to an aesthetic principle of economy: to yield the most truths from the least number of mathematical ideas. The principle of economy can also be seen in the fact that the fifth and final postulate—known as the parallel postulate—is not used until the 29th proposition, thus enabling Euclid to prove as many propositions as possible with an even more limited set of postulates. For example, Proposition 1 of Book I, shown in Table 2.1, "To construct an equilateral triangle on a given finite straight line," relies on two postulates, two definitions, and one axiom. This construction of the equilateral triangle using straightedge and compass was, for Euclid, prerequisite to its use in other theorems, and even to a discussion of its properties. Therefore, Euclid derived only the properties of objects that could be constructed by straightedge and compass. Table 2.1 presents Euclid's proof of Proposition 1 in such a way to highlight the axioms, definitions, and postulates he drew on, but this was not the format used by Euclid.

In addition to providing a treatise on geometric results that would be used as a textbook for over a thousand years, Euclid's *Elements* essentially defined the aims, methods, tools, and sequence of geometry in the western civilization. Thus when geometry began to be taught to schoolchildren in 19th-century American classrooms, it was the *Elements*, to a large extent, that first dictated most key aspects of the geometry curriculum:

Table 2.1. Two-Column Display of Euclid's Elements, Book I, Proposition 1

Proposition 1	
To construct an equilateral triangle on a given (finite) straight line.	
Let *AB* be the given finite straight line. It is required to construct an equilateral triangle on the straight line *AB*.	
Describe the circle *BCD* with center *A* and radius *AB*. Again, describe the circle *ACE* with center *B* and radius *BA*.	Postulate 3: To describe a circle with any center and radius.
Join the straight lines *CA* and *CB* from the point *C* at which the circles cut one another to the points *A* and *B*.	Postulate 1: To draw a segment from any point to any point.
Now, since the point *A* is the center of the circle *CDB,* therefore *AC* equals *AB*. Again, since the point *B* is the center of the circle *CAE,* therefore *BC* equals *BA*.	Definition 15: A circle is a plane figure contained by one line such that all the straight lines falling upon it from one point among those lying within the figure equal one another.
But *AC* was proved equal to *AB,* therefore each of the straight lines *AC* and *BC* equals *AB*. And things that equal the same thing also equal one another, therefore *AC* also equals *BC*.	Common notion 1: Things that equal the same thing also equal one another.
Therefore the three straight lines *AC, AB,* and *BC* equal one another.	
Therefore the triangle *ABC* is equilateral, and it has been constructed on the given finite straight line *AB*. Q.E.D.	Definition 20: Of trilateral figures, an equilateral triangle is that which has its three sides equal.

[a]Note that Euclid assumes here that the circles intersect in a point C; no axioms, postulates, or definitions guarantee this.

why students were to study geometry, *what* geometry they would study, and *how* they would see that geometry be developed.

THE IMPLICIT AIMS OF EUCLID'S GEOMETRY: *WHY?*

Given that Euclid's development of geometry was seen, for such a long time, as the finest example of a logical, deductive system, it is not surprising that teaching the *Elements* carried with it very theoretical, as opposed to practical, aims. In contrast, prior to the ancient Greeks, geometry's primary value was related to its practical applications in mensuration and in related fields of science. The Pythagoreans were the first to sever geometry from the needs of practical life, and they tended to treat geometry as a liberal art. After, Plato strengthened the severance by divorcing mechanics from geometry. And with Euclid, of course, not a single application can be found in his *Elements*; the value of geometry became enmeshed with a certain method—the deductive one—and with a certain form—the postulate-driven organization of knowledge.

The influence of Plato and Euclid can be seen in one strand of the four discourses that González and Herbst (2006) propose as being relevant to geometry education over the course of the 20th century: the discourse of formal argument maintains that the study of geometry provides students with the best access to logical reasoning. Compare this discourse with the three others, which emphasize both more mathematics and practical aspects of geometry learning: (1) the discourse of mathematical argument maintains that students should experience the work of *doing* mathematics, including making and proving conjectures; (2) the discourse of intuitive arguments involves aligning geometry with opportunities for students to model the world; and (3) the discourse of utilitarian argument maintains that geometry should provide tools for the future work of students (even those not pursuing careers in mathematics).

The discourse of formal argument has been inherited directly from Euclid's *Elements*. With no reference to how the propositions arise, or how they are proven, and with no connection to more utilitarian concerns or to intuitive understandings, the *Elements* privilege exclusively the discourse of formal argument. This discourse maintains that to study geometry through the *Elements* is to study the finest specimen of logical reasoning ever produced, and thus, to become familiar with an unsurpassed cultural achievement.

THE SCOPE AND SEQUENCE OF EUCLIDEAN GEOMETRY: *WHAT?*

Given the strict rules of logical dependence governing the *Elements*, it would have been unthinkable to Euclid for someone to study, for example, solid geometry (found in Book XI) before plane geometry, since the propositions of solid geometry depend heavily on those of plane geometry. Plane geometry occupies the first part of the *Elements* (Books 1–4, and 6), while solid geometry occupies the three last books (11–13). (Number theory occupies the other books.) This Euclidean order has had a lasting effect on the tradition of Western geometry. In particular, it has set up a standard in which all of plane geometry must be covered before embarking upon solid geometry—effectively barring many from studying solid geometry. The difficulty of the solid geometry axioms and theorems given by Euclid also contributes to making solid geometry seem almost unattainable to most students (and perhaps teachers).

The ordering of the Propositions contains other surprises for the uninitiated: one must wait until Proposition 46 of Book 1 to encounter the construction of a square, the most basic of geometric shapes, familiar to most 4-year-olds. The construction of the square appears so late because it depends on propositions about parallel and perpendicular lines, and about right angles and parallelograms. Thus, the *Elements* dictates a certain order that was motivated entirely by logic and economy, with no concern for pedagogical considerations. It should be noted that other Greek geometers, such as Archimedes, were much less bound to this Euclidean ordering, and, had Western geometry taken its cue from these geometers, the question of what to teach and when might have elicited very different answers, even at the earliest stages of geometry education.

The 13 Books also dictate a certain scope for geometry. So, while circles, triangles, similarity, and solids are well entrenched in the Euclidean tradition, other geometric topics such as conic sections—which were extensively studied by Euclid's contemporaries—have never been routine components of the geometry syllabus. Given the logical tightness of the *Elements*, mathematics educators have had to appeal to the other discourses described by González and Herbst (2006) in order to include new topics in the syllabus or to change the order of existing topics. So, for example, if a teacher wanted to treat solid geometry alongside plane geometry (instead of after) she might have to appeal to a different discourse—perhaps arguing that such a move would be more intuitive to learners, given that solid geometry describes more directly students' familiar, three-dimensional world. In fact, in the mid-twentieth century, educators began to express increased appreciation for the pedagogical value of analogy, comparison, and experience, thus arguing for an integrated approach to the teaching of plane and solid geometry. They argued that students might appreciate results in plane geometry (such as the sum of the measures of the angles of a triangle) more when they can also experience them in solid geometry. Euclid would probably have objected to such an argument, fearing the compromising effects on the clarity and consistency of his logical system.

THE METHODS AND INSTRUMENTS OF EUCLIDEAN GEOMETRY: *HOW?*

In his 1909 thesis titled *A History of the Teaching of Elementary Geometry*, Stamper singled out four "methods of attack" invented and used by the ancient Greek geometers: analysis, *reductio ad absurdum* (proof by contradiction), exhaustions, and loci. The role and importance of these methods of attack have changed substantially—so much so that most high school teachers would have a hard time finding what unites them as geometric ideas. In fact, they are quite diverse: both the method of exhaustion and *reductio ad absurdum* were used as methods for proving results, while the method of analysis was used by ancient Greek geometers to lead to discovery. And loci played both discovery and proving roles,[1] even though today they are treated quite sparingly, and almost as a separate topic in geometry.[2]

Three of the four methods of attack were used extensively in the *Elements*—all but loci. The method of exhaustion was perfected by Eudoxus, but most famously put to use by Archimedes; Euclid makes extensive use of it in his 12th book to prove that the areas of circles are to one another as the squares of their diameters and that the volumes of spheres are to one another as the cubes of their diameters. Proof by *reductio ad absurdum* is frequently illustrated nowadays using the Euclidean proof of the irrationality of $\sqrt{2}$. The technique involves a type of logical argument where one assumes a claim, the negation of what one wishes to prove true, then arrives at a contradiction of a known result, and then concludes the original assumption must have been wrong, since it gave rise to an "absurd" result. In contrast, the method of analysis involves supposing for the time being that the desired theorem is true or the sought problem is solved so as to find the necessary underlying conditions. After the necessary conditions are found, the construction is made, the deductive proof is given, and then, for the ancient Greeks, a discussion (the *diorismus*) follows regarding the conditions under which the problem is or is not solvable. Despite its importance, the analysis never appears in the final presentation of the proofs offered by Euclid.

While methods are important in all fields of mathematics, and often take on distinct flavors, no other field of mathematics has been so ruled by its physical tools as geometry has been: the compass and straightedge virtually defined the boundaries of Euclid's *Elements*. In Proposition I given above, the compass is used to draw a circle and the straightedge to draw line segments, and no other tools appear, not a ruler or protractor, and none of the mechanical instruments that Eudoxus and Menæchmus used. In fact, for this reason, conic sections, which had already been so extensively studied by other ancient Greek geometers, are absent from the *Elements*.[3] One can appreciate the impact of Euclidean instruments by considering how different the *Elements* may have looked had measurement tools such as the ruler or the protractor been admitted by Euclid. For example, consider Proposition 2, which involves constructing a segment from a given point that is equal in length to a given segment. A ruler would have made this proposition superfluous.

Over the course of a thousand years, Euclid's *Elements* gained increased prestige and canonicity, as scholars around the world studied, transcribed, and translated them. But they were studied primarily by men, and primarily by men who had devoted their lives to the study of geometry. They were not written with pedagogical purposes in mind! Nevertheless, they would become the basis of geometry education in the United States, offering universities first, and then schools, an initial course of study. Many tensions arose in response to the use of Euclidean geometry, especially with respect to its theoretical, formal orientation. For example, with no practical applications, the *Elements* were strikingly not well suited to meet the growing utilitarian needs of the geometry curriculum, which had to cater to an audience much larger and more diverse than the one consisting of privileged, mature, male scholars.

With this initial framing in mind, the next section jumps ahead two millennia to the beginning of geometry education in the United States. There will be several opportunities to revisit some of the initial constraints and deficiencies inherent in Euclid's *Elements*, and to examine their lasting influence on the development of the geometry curriculum.

NOTES

1. Loci were also used to prove theorems, such as the concurrency of the perpendicular bisectors of a triangle ABC. Here, the locus of the points equidistant from A and B, and the locus of points equidistant from points B and C, can be shown to intersect. Since the point of intersection is equidistant from A, B, and C, it is also on the perpendicular bisector of AC. Therefore, the three bisectors are concurrent.
2. A geometric locus is defined as the set of all points that satisfy a given set of conditions. For example, a circle can be thought of as a locus in that it is the set of all points that are equidistant from a given point, the center. Similarly, the perpendicular bisector of a segment AB is the locus of all points equidistant from A and B.
3. Envelopes of tangent lines to conic sections can be constructed by straightedge and compass, but these were not included by Euclid.

CHAPTER 3

STEPS IN THE HISTORY OF THE GEOMETRY CURRICULUM

In looking back through the history of the geometry curriculum in the United States, one is struck by its relative constancy, particularly with respect to the dramatic changes that have occurred in other areas of society, and even education. One is also struck by the variety of influences that have nudged the geometry curriculum in certain directions—small and slow-moving as those nudges may be. For instance, we will see that the appointment of Claude Crozet in 1817 to the West Point Military Academy began an invasion of French mathematics: The geometry textbook of Adrien-Marie Legendre—and textbooks were the defining curriculum then—began taking the place of Euclid at the American universities, and the influence of the British waned. Naturally, policy decisions were *at times* also an important source of nudging, such as on the occasion of the College Entrance Board Examination report of 1959, which, arriving post-Sputnik, encountered an unusually receptive audience. While it is unwise to trace changes back to single events, such as academic appointments or policy reports, such events can often provide a useful way of analyzing and assessing the development of the intended geometry curriculum—and of setting up signposts that can structure a historical report such as this. Figure 3.1 provides a brief graphic timeline of the notable events identified in the monograph. To begin, I will paint a picture of the state of the geometry curriculum in the United States prior to the middle of the nineteenth century.

The History of the Geometry Curriculum in the United States
pp. 13–90

1844: Geometry lands on the list of university entrance requirements
1850s: Early pedagogical influences on the geometry curriculum

1860s: A mathematical challenge to the dominance of Euclidean geometry

1892: First attempts to standardize the school geometry curriculum

1902: Perry and Moore and theories of learning

1920s and 1930s: Committees and their Reports Redux

1935: The Bourbaki Group and its impact on geometry

1944: Post-War developments

1955: Beginning of the "era of reform"
1960: Time for transformations

1975: The NACOME Report redefines "basic skills"
1980s: The De-degradation of "Geometric Consciousness"

1989 NCTM: *Curriculum and Evaluation Standards for School Mathematics*

1991: Dynamic Geometry Software

2000: New and Improved Principles and Standards

Figure 3.1. Notable events timeline.

THE GEOMETRY CURRICULUM IN THE UNITED STATES IN THE EARLY NINETEENTH CENTURY

Until the middle of the nineteenth century, geometry was generally taught only in the universities, as was the case in Europe. In fact, little mathematics beyond arithmetic was taught in the pre-college schools. At the close of the eighteenth century, the only mathematics needed for admission to college included the rules and processes of arithmetic. In the colleges, Stamper (1909) reports that it was only around 1725 that geometry became an integral part of mathematics courses. Naturally, Euclid's *Elements*, or something similar, was the text used by universities such as Harvard and Yale, and it was taught in the second or third year. Typically, the geometry curriculum would include the first six books of Euclid, and perhaps, as was the case at the University of Pennsylvania, Books XI and XII.

Despite the theoretical orientation of the *Elements*, the primary aims of teaching geometry during this time were supposedly for its practical and applicative characteristics. As Smith and Ginsberg (1934) reported, geometry was seen as being "merely an aid to the study of astronomy" (p. 13). This statement exemplifies the first modern instance of what González and Herbst (2006) called the *intuitive* discourse, in which geometry is seen as providing tools to help students model the world.

By 1820, there were 14 geometry textbooks "on the market," all modeled on Euclid's formal approach. These included Robert Simson's *The Elements of Euclid* (1781) and John Playfair's *Elements of Geometry* (1795). A page from Simson's book is shown in Figure 3.2; it depicts Proposition 48, which proves what is arguably the most famous result in geometry, the Pythagorean theorem. In the right margin, Simson supplied the various definitions and postulates required along the way. The diagram illustrating the theorem, along with the labels for the points, figures prominently, as it did in Euclid's text. As we shall see, this reliance on diagrams would later prove to be problematic for early twentieth-century mathematicians, who began to suspect diagrams as potentially misleading.

The *modus operandus* in the classrooms at this time echoed the *Elements'* lack of explanation or interpretation, and consisted of memorizing the definitions, axioms, and propositions provided in the text. The textbooks used contained no exercises, that is, no opportunities for students to do applied or original work; for example, to apply the method of *reductio ad absurdum* to another proposition. The goal was for students to learn and appreciate the work of Euclid; Euclid had laid out the text of geometry and there was little reason to change or extend it. The best one could do was to commit it to memory.

However, when Claude Crozet was hired at the West Point Military Academy in 1817, he brought with him the French influence in the form of a new textbook by Adrien-Marie Legendre, which had been published in 1794, and first translated into English in 1819 by Charles Davies. Prior to this, the textbooks used in American colleges were mostly written in England. Since the curriculum *was* literally the textbook, with university students memorizing one proposition after another in the given sequence, any change in textbook was momentous: providing an alternative to Euclid was tantamount to changing the very definition of geome-

OF EUCLID. 51

Book I.

PROP. XLVII. THEOR.

IN any right angled triangle, the ſquare which is deſcribed upon the ſide ſubtending the right angle, is equal to the ſquares deſcribed upon the ſides which contain the right angle.

Let ABC be a right angled triangle having the right angle BAC; the ſquare deſcribed upon the ſide BC, ſhall be equal to the ſquares deſcribed upon BA, AC.

On BC deſcribe [a] the ſquare BDEC, and on BA, AC the ſquares GB, HC; and thro' A draw [b] AL parallel to BD or CE, and join AD, FC. therefor becauſe each of the angles BAC, BAG is a right angle[c], the two ſtraight lines AC, AG upon the oppoſite ſides of AB, make with it at the point A the adjacent angles equal to two right angles; therefor CA is in the ſame ſtraight line [d] with AG. by the ſame reaſon, AB and AH are in the ſame ſtraight line. and becauſe the angle DBC is equal to the angle FBA, for each of them is a right angle, add to each of them the angle ABC, and the whole angle DBA ſhall be equal [e] to the whole FBC. and becauſe the two ſides AB, BD are equal to the two FB, BC, each to each, and the angle DBA equal to the angle FBC; the baſe AD ſhall be equal [f] to the baſe FC, and the triangle ABD to the triangle FBC. now the parallelogram BL is double [g] of the

a. 46. 1.
b. 31. 1.
c. 30. Def.
d. 14. 1.
e. 2. Ax.
f. 4. 1.
g. 41. 1.

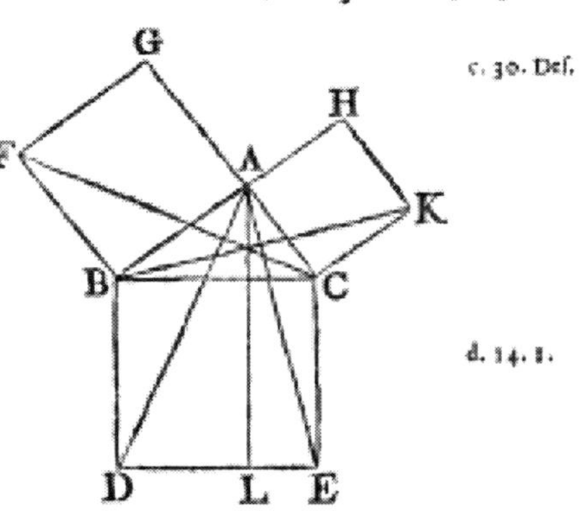

G 2

Figure 3.2. The proof of the Pythagorean theorem in Simson's version of Euclid's *Elements*.

try, and opened the door to subsequent alterations in the scope and sequence of a geometry course of study.

Although many mathematicians had already identified some of the logical difficulties in Euclid's *Elements*, and proposed alternatives, Legendre's efforts appeared to have the greatest and longest-lasting impact. His treatment of the *Elements* abandoned the sequence of Euclid and thus dramatically simplified the subject matter, while maintaining its logical soundness—thereby ensuring its acceptance by the mathematical world. For example, Legendre decided to admit hypothetical constructions, such as Proposition 1, thereby challenging the necessity Euclid felt to ascertain constructions no matter how abstruse and unmotivated they were. Future mathematics educators would find Legendre's reorganization helpful to meeting the psychological challenges of teaching geometry.

In addition to reorganizing Euclid's material, Legendre introduced measure into geometry. Euclid developed his geometry without assigning numerical measurements to line segments, angles, or areas. Instead, he had used the undefined concept of equality (congruence) for line segments, which was established by first placing one segment on the other and then determining whether they coincided exactly—a method that became known as superposition. In this way, the equality or inequality of line segments was to be perceived directly from the geometry, without the assistance of real numbers to measure. Legendre instead assumed correspondence between a line segment and a number, namely, its length, allowing him to measure numerical quantities, rather than rely on comparisons of one segment to another. He introduced the concept of number by defining the concept of a unity: "If *A; B; C; D* are lines [line segments], one can imagine that one of these lines, or a fifth, if one likes, serves as a common measure and is taken as unity. Then *A; B; C; D* represent each a certain number of unities, whole or fractional, commensurable or incommensurable, [...]" (1823, p. 61). While for Euclid, geometry was a matter of shape and proportion; Legendre turned it in to a matter of numerical quantities also. This paved the way—albeit tentatively—for students, teachers, and textbook writers to see geometry as an integrated part of mathematics—connected to other domains of mathematics with which they were familiar—rather than a self-contained, self-sufficient discipline of its own.[1]

Despite Legendre's many improvements, as acknowledged by mathematicians at the time, mathematics educators did not seem to accept readily the notion that Euclid's *Elements* contained many logical and mathematical shortcomings.[2] One shortcoming was the absence of undefined terms in the *Elements*, such as that of equality. Unlike his predecessor Aristotle, Euclid did not recognize that every deductive system must begin with undefined terms from which all others are defined.[3] Another shortcoming was the incompleteness of Euclid's postulate system; Euclid unknowingly assumed certain facts that did not appear in his list of postulates or axioms. For instance, he used the notions of "betweenness" and "continuity" without postulating them. Finally, Euclid subscribed to a problematic (and complex) view of logical truth, in which postulates should be *obviously* true instead of *assumed* true (complex in the sense that Euclid wavered about the fifth postulate, which was perhaps not as obvious to him as he would have liked). Fortunately, geometers were clever in finding ways of addressing these shortcomings while still preserving much of the flavor and content of Euclid's *Elements*. To this end, in 1898–1899, David Hilbert developed a set of axioms for Euclidean geometry that were extremely influential (see Eves, 1972, for a list comparing Euclid's and Hilbert's axiom systems).

1844: GEOMETRY LANDS ON THE LIST OF UNIVERSITY ENTRANCE REQUIREMENTS

The influence of university courses on the high school geometry curriculum began very early on, and has continued to current times. When universities, in 1844, began to place geometry on the list of entrance requirements, high schools were required to assume the teaching of the subject. Naturally, they adopted the curriculum articulated by the universities, namely, Euclidean geometry, as well as the implicit goals. In the *Era of the Text*, as Herbst (2002) called this period of time, studying geometry was equivalent to mastering the Euclidean body of knowledge as presented in a particular text—whether it be Simson's, Playfair's, or Legendre's. While these texts differed in some ways, as discussed earlier, they shared the aesthetic goal of providing maximal results from a minimal number of axioms, regardless of the effect on the difficulty of the proof—and thus the intelligibility for students. For instance, if a theorem could be proved without the parallel postulate, it was considered more valuable for students to know, even if it doubled the difficulty and length of the proof. [4]

Stamper (1909) reported that the textbook played a primary role in high schools at the time, and that, as a result, the tendency of the students was to learn theorems, definitions, and proofs by heart. Indeed, there were no general techniques ever given for explaining the origins of a proof, each proof being instead viewed as especially suited to the proposition in question. While some textbooks were written specifically for high school students, they bore a strong resemblance to the pattern of formalism present in the college-level texts. Very little attention was devoted to the problem of motivating students; and, it was generally assumed that by virtue of memorizing the propositions in their correct order, students would be able to transfer smoothly their ability to reason logically to other domains of life. Students were not expected to actually use the tools of geometry favored by Euclid—the straightedge and compass—nor would they be expected to apply the methods of geometry to novel situations; geometry was a body of knowledge to be memorized.

The long-standing trend of simply pushing mathematical content down to increasingly younger students evidenced at this particular juncture has persisted today, but at that time it was exacerbated by the initial lack of concern for developmental appropriateness. It must be remembered, however, that by 1860, there were only 40 public high schools in the United States (Stamper, 1909), which meant that the audience of high school geometry was quite limited, and likely to be college-bound—perhaps closer in terms of privilege to the ancient Greek scholars than any audience since.

Though little changed in the geometry curriculum, or in the assumed goals of geometry education, the university-sanctioned transfer of college geometry to the high school setting established a precedent in the American geometry curriculum that would prove to be very difficult to alter: the text had been handed down, and expectations had begun to form for both teachers and learners (who later become parents and administrators—as well as teachers!) about the *what, how,* and *why* of geometry education.

1850s: EARLY PEDAGOGICAL INFLUENCES ON THE GEOMETRY CURRICULUM

The mid-nineteenth century marked the first pedagogically oriented influence in the teaching of geometry. Working in Switzerland, Johann Pestalozzi was greatly influenced by the ideas of philosopher Jean-Jacques Rousseau and insisted that students learn through activities and concrete objects—rather than through words—and that they should be free to pursue their own interests and draw their own conclusions. For the teaching of geometry, this "method" translated into an increased emphasis on including inductive procedures in teaching, procedures that were very foreign to Euclid's carefully devised methods of analysis and synthesis. Pedagogical concerns, not mathematical ones, drove Pestalozzi's methods. This may help explain why Pestalozzi's influence on American schooling was relatively minimal. Had Archimedes' work formed a most important part of the geometry canon, things might have been different; some of the greatest achievements in ancient Greek geometry were fueled by Archimedes's extensive use of inductive procedures, as well as more "mechanical" methods. For example, Figure 3.3 shows an Archimedean spiral, which Archimedes used to square the circle (an impossible task for the straightedge and compass), by relaxing the strict limitations of

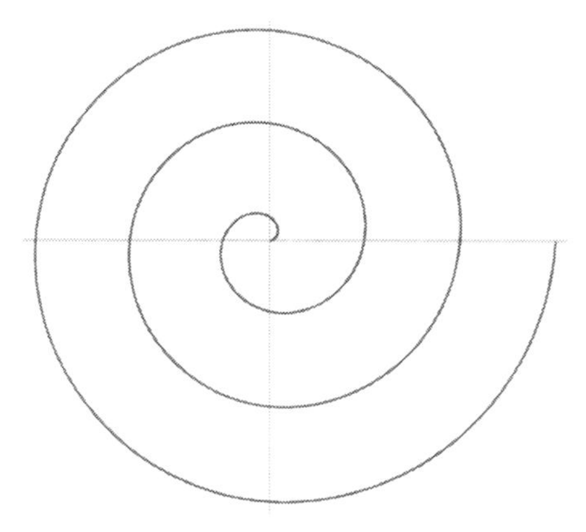

Figure 3.2. Three turns of one arm of an Archimedean spiral.

Euclid's geometry. We would now represent this curve in polar coordinates as $r = a + b\theta$.

But given the prominence of Euclid, the Pestalozzian-inspired methods were almost intrinsically antithetical to the geometry of the time. This seems to be a common theme in the history of the geometry curriculum, namely, that new methods or aims or instruments are introduced without due attention to the suitability of the changes to the variables of the curriculum implementation that remain constant. Eventually, as Wren and McDonough (1934b) noted: "Curricula were organized and expanded rapidly with no particular plan or definite educational objective in view" (p. 218).

It is unclear whether the Pestalozzian influence extended to colleges, but the emerging idea for schools suggested that there should be less emphasis on rigorous proofs, and that students should be encouraged to work out exercises on their own. Wren and McDonough (1934a) reported that the gradual change from dogmatic methods of teaching to more inductive (or "intuitive," as they would later be called) ones did not become widespread until the first half of the twentieth century. In the meantime, textbook titles such as *First Lessons in Geometry* (Hill, 1854), *Inventional Geometry* (Spencer, 1876), *Observational Geometry* (Campbell, 1899), and *Constructive Geometry* (Hedrick, 1917) began to emerge, indicating that the time was ripe for change. In fact, already in the preface to Hill's *Geometry for Beginners* (1887), we find the following description: "I have, therefore, avoided reasoning, and simply given interesting geometrical facts, fitted, I hope, to arouse a child to the observation of phenomena, and to the perception of forms as real entities" (quoted in Reeve, 1930, p. 10). By easing the way for descriptions of geometry such as "observational," "inventional," and "concrete," the Pestalozzian influence may have helped prepare the ground for the emergence of the nondemonstrative junior high school geometry that would evolve over a half-century later.

Inventional Geometry (1876) was a particularly interesting book, being the most structurally different than the others. It also took very seriously the idea that students should construct their own mathematics, and it emphasized the role geometry could play in providing insightful visualizations. Written by William George Spencer (whose son Herbert Spencer, an esteemed philosopher and educator, wrote the introduction), this "little book" consisted of 80 pages of geometric problems, and was intended to introduce "the beginner to geometry by putting him at work on problems which will not only thoroughly familiarize his mind with geometrical ideas, but will exercise, at the same time, his inventive and constructive faculties—a kind of mental practice of much importance, but generally neglected in our schools" (p. 3). In describing the strength of the book, Spencer articulates some of the goals of learning geometry: "It educates the hand to dexterity and neatness, the eye to accuracy of perception, and the judgment to the appreciation of beautiful forms" (p. 7). These goals were achieved by having the student actually construct *all* the shapes and configurations Spencer could imagine, in contrast with most other textbooks at the time:

> Instead of dictating to the pupil how to construct a geometrical figure—say a square—and letting him rest satisfied with being able to construct one from that dictation, the author has so organized these questions that by doing justice to each in its turn, the pupil finds that, when he comes to it, he can construct a square without aid. (pp. 8–9)

Spencer's book began with solid geometry—another important transgression of Euclidean geometry—yet constantly fuses solid and plane geometry questions and ideas. The first question given to the student, after a limited introduction, illustrates Spencer's belief that students need concrete experiences with definitions: "Place a cube with one face flat on a table, and with another face toward you, and say which dimension you consider to be the thickness, which the breadth, and which the length" (p. 16). Some questions were impossible, in that they have no answer, while others seemed to require much more imaginative thinking than traditional exercises; for instance: "7. Can two lines meet together without being in the same plane?" (p. 18). The triangle does not appear until page 26, and in this cloaked form: "44. Make a linear figure having the fewest boundaries possible, and in it write its name, and say why such figure claims that name." Many of the questions involved having the students construct the objects being defined, and a surprising number involved various forms of dissection, such as this one: "150. Can you invent a method of dividing a circle into four equal and similar parts, having other boundaries rather than the radii?" (p. 45). In addition to fusing plane and solid geometry, Spencer regularly incorporated arithmetic in his questions; the following one illustrates his belief in the role geometric images could play in other domains of mathematics: "294. Exhibit to the eye that 1/2 + 1/3 + 1/6 = 1" (p. 69).

In between the traditional textbooks characterizing the "Era of the Text," and the rather progressive ones mentioned above, a few other middle-ground textbooks were published, including Greenleaf's (1858) *Elements of Geometry with Practical Applications to Mensuration* and Chauvenet's (1887/1898) *A Treatise on Elementary Geometry with Appendices containing a Collection of Exercises and an Introduction to Modern Geometry*. In these textbooks, students were given opportunities to craft proofs for "original" propositions—which often required some ingenuity. Herbst (2002) signaled the emergence of such textbooks as the *Era of Originals*, during which textbooks expected students to not only learn the theorems of Euclidean geometry, but to apply the methods used in proving those theorems to novel situations. Chauvenet motivated the presence of "originals" in his textbook by warning that the "power of grasping and proving a simple geometrical proof" can "never be gained by memorizing demonstrations" (Chauvenet, 1887, p. 5). Later, Young (1906) would write that students could "only learn to demonstrate by demonstrating" (p. 259).

The use of "originals" presumed, as Herbst has pointed out, that students would learn to reason by reasoning themselves, instead of relying on learning by example. Of course, "originals" were *always* about proof, and not about doing the original kind of thinking that would be required to find solutions to the questions given in Spencer's textbook, or the problem-solving questions that would appear in later twentieth-century textbooks. However, as aims shifted and expectations changed, "originals" slowly turned into "exercises," which were, as Wentworth (1888) described them, "not so difficult ... but well adapted to afford an effectual test of degree in which [the student] is mastering the subject of his reading" (p. iv). The number of these exercises, their placement in the text, and their level of difficulty were altered and debated in the ensuing decades, but remained a permanent fixture in geometry textbooks. Nonetheless, their intent shifted over the years as they went from aspiring to Chauvenet's ideal of providing students with opportunities to develop new ideas pertaining to the study of geometry to fulfilling Wentworth's goal of providing the student with opportunities to practice what had already been learned. In other words, the aims of geometry did not much change; instead, the means of teaching it did.

1860s: A MATHEMATICAL CHALLENGE TO THE DOMINANCE OF EUCLIDEAN GEOMETRY

For many centuries, geometers had tried in vain to rid Euclid's *Elements* of the fifth postulate, the one that Playfair had formulated as follows: Through a given point not on a straight line there is exactly one line through the given point parallel to the given line. Euclid himself was uncomfortable with this postulate, and wished he could find a way to show it depended on the first four. When geometers finally came to the realization that the fifth postulate was in fact independent of the first four, and, more, that negating it could result in the creation of entirely new and seemingly wild geometries, the effect on geometers and philosophers alike was profound. The development of non-Euclidean geometries—geometries that replaced Euclid's fifth postulate with one of its negations—destabilized many nineteenth-century philosophers, who saw Euclid's geometry[5] as the description of the fundamental properties of spatial reality. All of a sudden, the angles of a triangle could sum to less than 180° (or even more than 180°). These triangles, of course, didn't live on the plane; instead, they lived on other surfaces such as spheres. Figure 3.4 compares a Euclidean triangle with a non-Euclidean triangle constructed on the Poincaré disk, which is a model of a geometry in which there can be many lines through the given point parallel to the given line; the angles of the hyperbolic triangle sum to less than 180°.

The development of non-Euclidean geometries also altered the established views concerning the nature of mathematical truth, and the extent to which Euclidean geometry could continue to act as the paradigm of truth. As Kline (1953) wrote, "though the logical exercise [of erecting new

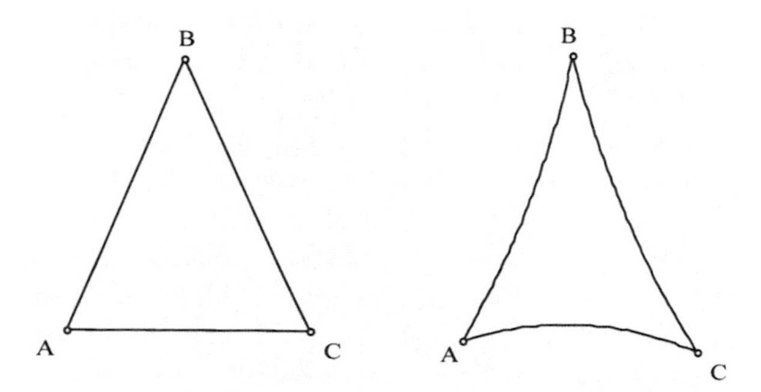

Figure 3.4. Comparing Euclidean and non-Euclidean triangles.

geometries] was over the conclusion lingered on in people's minds: *There are geometries different from Euclid's*" (p. 417, *italics in original*).

During the 1860s, just as mathematicians were rediscovering Bólyai's and Lobachevskii's first tentative work on non-Euclidean geometry, two new non-Euclidean works by Riemann and Helmholtz appeared on the scene. Elliptical geometry, for example, negates the fifth postulate by asserting that *no* line can be drawn through a point parallel to a given line. The surface of a hemisphere[6] provides a good model of this geometry: if lines are taken to be the great circles of the sphere (like the longitudes of the earth), then, indeed, there exist no parallel lines. In this geometry, the sum of the angles of a triangle will be greater than 180°.

Soon after, by the 1870s, the projective geometry that had been developed in France in the first decades of the nineteenth century had finally been recognized in England, and hailed as "the modern geometry." One way of thinking of projective geometry is to consider its tools of construction: one abandons the compass and works solely with the straightedge. Thus, the basic elements of projective geometry are points, lines, and planes. This means that many of the geometric objects that are important in Euclidean geometry, such as circles, parallel lines, angles, equilateral triangles, and squares, are not relevant to projective geometry. Another way of getting a sense of projective geometry is to think of familiar types of projections found in art and map-making: projections used to create paintings that appear to be in perspective and projections of the earth on two-dimensional surfaces. Both examples illustrate how ratios, angles, and lengths are not invariant under projective transformations.

The work of Felix Klein (1849–1925) was highly influential in bringing all of the geometries together under one umbrella. He first showed that it was possible to consider both Euclidean and non-Euclidean geometries as special cases of a projective surface. This led to an important corollary, namely, that non-Euclidean geometry is consistent if and only if Euclidean geometry is consistent. Now, the controversial status of non-Euclidean geometry could be overcome, and achieve equal footing with Euclidean geometry.

Projective geometry essentially subsumes Euclidean geometry, since Euclidean geometry can be shown to be a special case of projective geometry. In addition, projective geometry leads naturally to the dissolution, for mathematicians, of the traditional distinction between plane and solid geometry, that is, it encourages the *fusion* of plane and solid geometry. Already, Gergonne had suggested, in 1826, that projective geometry should lead to a rethinking of the ancient Greek division. For, as French mathematician Chasles (1937/1989) wrote, projective geometry makes evident the "intimate and systematic connection between three-dimensional shapes and two-dimensional ones" (my translation, p. 191).

French and Italian mathematics educators were more enthusiastic about the possibility of that fusion than their American counterparts (Bkouche, 2003). Nonetheless, this period of time is important even for geometry education in the United States since the idea of fusion between plane and solid geometry had to be well accepted by mathematicians—who had, at the behest of Euclid, previously argued against fusion—before it could begin making inroads in school geometry education. Of course, school geometry did not then, and has still not now, become more projective in nature.

The result of the discovery of non-Euclidean geometries, as Richards (1988) observed, with particular attention to mathematics education, was that two communities of mathematicians began to conflict. On the one hand, there was the emerging nineteenth-century mathematical community influenced by the changes in subject matter and, on the other, there were the more traditional mathematicians and educators who conceived mathematics as a subject whose primary purpose was to discipline and train the mind (and who believed that only Euclidean geometry was best suited for these purposes). Some early twentieth-century mathematics educators argued for the benefits of including non-Euclidean geometries in the curriculum, citing, for instance, the importance of teaching students a more current, "alive" geometry. But their recommendations fell on deaf ears in the sense that the dominant discourses for geometry over a large part of the twentieth century focused on logical reasoning and utilitarian arguments, and not on arguments that might have been more hospitable to purported benefits of teaching new geometries.

Nevertheless, the advent of non-Euclidean geometries did have two strong influences on the geometry curriculum. First, with the discovery of non-Euclidean geometries, the *Elements* began to be more carefully scrutinized by mathematicians, resulting in the discovery of several logical omissions. These, as we shall see, were eventually seized upon by mathematics educators, who became interested in finding tactics that could rescue the logic without compromising the status of the *Elements.* Second, with the mathematical and philosophical doubts now surrounding geometry, many mathematicians turned to other areas of mathematics for foundational texts that could restore and perhaps even surpass the standards of rigor established by the ancient Greeks. This eventually prompted the decision to found the field of analysis—an exploding and vast area of mathematics—on arithmetic instead of geometry, and would set the scene for the emergence of the Bourbaki group in the 1930s. The impact of this turn of attention to numbers, and away from geometry, lasted well into the next century, and continues today.

1892: FIRST ATTEMPTS TO STANDARDIZE THE SCHOOL GEOMETRY CURRICULUM

This section discusses a span of 20 years, during which three important national committees—the first of their kind in the United States—were formed, in addition to one international commission: the Committee of Ten, Committee on College Entrance Examinations, the Committee of Fifteen, and the International Commission on the Teaching of Mathematics (ICTM, now known as ICME). The period of time during which these committees worked signaled the first attempts to influence, define, and systematize the geometry curriculum in American schools.

In 1892, in an effort to change school practices on a widespread basis, the National Education Association (NEA) appointed 10 committees to examine the secondary school curriculum. These became known as the Committee of Ten; their main challenge, according to Ravitch (2000), was to accommodate the diversity of available studies into an academic curriculum that was appropriate for all students. One of the committees was devoted to the teaching of mathematics—which, significantly, treated geometry as a valuable part of mathematics that *all* secondary students should study. While the physical sciences could train students in inductive reasoning, the committee subscribed to the view that formal geometry provided "the best possible arena for training in deductive reasoning."

Although the report of the Committee in 1894 had considerable influence, Taylor (1930) commented 37 years later that "I have read the report more than once and have wondered how long after my time it will be until American high schools shall have moved up to it" (p. 228). On the other hand, Sizer (1964) thought that the report did have considerable influence both on school policy and on the thinking of educators. In fact, it also strongly impacted the shape of textbooks, albeit in subtle ways not necessarily consonant with the loftier recommendations hinted at by Taylor. Furthermore, the report certainly influenced the pre–high school geometry curriculum, where there was much less traditional influence with which to contend. It also brought attention to some issues that continued to be relevant in the century that followed, particularly with respect to the degree of student-centered learning that geometry could accommodate and to the compartmentalization between geometry and algebra.

Introducing Geometry in the Junior High School

Instead of providing specific models that could be used by textbook authors or teachers, the committee restricted itself to issuing general suggestions about organization and presentation. The most fundamental of these suggestions—one that was to have discernable impact on the teaching of geometry in schools, albeit not until the turn of the century—was to advocate the introduction of "concrete" geometry throughout elementary school, to be followed by more formal work in "demonstrative" geometry in the high school. The statements made in the recommendations prefigured policy recommendations as modern as the National Council of Teachers of Mathematics (NCTM) *Principles and Standards for School Mathematics* (2000), with respect to communication and modeling. In addition, the committee encouraged the decompartmentalization of geometry by advocating that teachers and textbooks connect geometry not only with arithmetic, but also with physics.

> In regard to the teaching of concrete geometry, the Conference urge that while the student's geometrical education should begin in the kindergarten, or at latest in the primary school, systematic instruction in concrete or experimental geometry should begin at about the age of ten for the average students, and should occupy about one school hour a week for at least three years. From the outset of this course, the pupil should be required to express himself verbally as well as by drawing and modeling. He should learn to estimate by the eye, and to measure with some degree of accuracy, lengths, angular magnitudes, and areas: to make accurate plans from his own measurements and estimates; and to make models of simple geometric solids. The whole work in concrete geometry will connect itself on the one side with the work in arithmetic, and on the other with elementary instruction in physics. (NEA, 1894, p. 24)

The rise of junior high schools after 1910 provided an excellent opportunity for this committee's recommendations about concrete geometry to take root, since many high schools were reluctant to abandon demonstrative geometry, and many educators saw junior high school as the optimal time for preformal work. Quast (1968) notes that the claim that intuitive geometry is valuable to later work was left implicit in these recommendations, for in terms of the goals and methods used in the teaching of demonstrative geometry, the two subjects were very different. In fact, it would be many years before anyone was able to articulate the value more explicitly. One might even argue that explaining and supporting the connection between intuitive and formal ways of knowing in mathematics continues to challenge mathematics education researchers today. Also interesting is the deliberate separation of "concrete" or "experimental" geometry from "demonstrative" geometry—a hard distinction that is often softened in contemporary policy documents or textbooks.

Recommendations: Exercises, Decompartmentalization, and Fusion

The discussion of "concrete" geometry was relatively short compared to that for demonstrative geometry, where the committee saw more need to systemize the confused practices at the time. Here, however, the report was less ambitious in that it followed the traditional formal sequence of topics of Euclidean geometry (including the exclusion of intuitive methods, constructions, and experimentation), and even virtually ignored any applications of geometry, which were once seen as integral to the goals of teaching geometry. The committee continued to assume that mathematics had a general disciplinary value, which freed them from calling into question the traditional content of the courses. In fact, they argued that the logic of geometric truths were more important than the truths themselves, thus narrowing the objectives of geometry to "the art of demonstration." The refocusing instigated by the committee led to the *Era of Exercise* (Herbst, 2002), during which geometry texts endeavored to devise ways of helping students learn to *do* proofs. Incidentally, it is interesting to read the committee's statement regarding the goals of geometry, particularly the emphasis on the aesthetic characteristics of formal geometry, which grew in importance as "the art of demonstration" took center stage:

> [The report] insists on the importance of elegance and finish in geometrical demonstration, for the reason that the discipline for which geometrical demonstration is to be chiefly prized is a discipline in complete, exact, and logical statement. If slovenliness of expression, or awkwardness of form is tolerated, this admirable discipline is lost. The Conference therefore recommends an abundance of oral exercises in geometry—for which there is no substitute—and the rejection of all demonstrations which are not exact and formally perfect. (pp. 24–25)

Of course, the student can "begin to devise constructions and demonstrations for himself" only *after* he or she has "acquired the art of rigorous demonstration" (p. 25). This recommendation reflected a pedagogical belief that continues to influence school mathematics, namely, that students must first learn the required techniques and forms before being able to engage in more creative or independent mathematical activity.

The committee's emphasis on the art of demonstration—on students' ability to *do* proofs—led to several changes, particularly in textbooks, having to do with enabling and supporting students' performances in proof "exercises." For example, textbooks began instructing students on methods and strategies for coming up with proofs. In fact, the dominant textbooks at the time, Wentworth's *A Text-book of Geometry* (1888) and

Wentworth and Smith's *Plane Geometry* (1910) were popular, in part, because of the abundance of "exercises" they included (Quast, 1968). These "exercises" differed in style and aim from the "originals" discussed above as they were intended to provide practice in "the art of demonstration" rather than opportunities for original thought. George A. Wentworth, who studied under Benjamin Peirce at Harvard, was a major figure in the textbook market for both school algebra and geometry. Wentworth's son had collaborated with David Eugene Smith to revise his father's textbooks. Smith was a member of the Committee of Fifteen and chair of the American Commission of the International Commission on the Teaching of Mathematics. In his book, Wentworth (1892) criticized the practice of having students memorize proofs of theorems as "a useless and pernicious labor" (p. iv).

In addition to advocating the use of abundant (but "not so difficult") exercises, Wentworth drew attention to the importance of language use in geometry, and to the wide gap separating students' natural language with the formal language of geometry. He recommended that teachers permit students to "write [their] proofs on the blackboard in [their] own language," taking care that "[their] language be simplest possible, that the arrangement of work be vertical (without side work), and that the figures be accurately constructed" (p. v). Beginning with his 1888 textbook, Wentworth also showed concern for the format of geometric demonstrations. For example, he displayed every theorem and its proof—along with its accompanying figure—in a single page. He also introduced conventions such as providing reasons for *each* step, which were indicated in small italic type between that step and the one following, and having each distinct assertion begin a new line. The formatting changes he introduced were all intended to facilitate student understanding: the pupil "acquires facility in simple and accurate expression" and "rapidly *learns to reason*" (p. iv, *italics in original*). Wentworth thus inaugurated a trend toward breaking proofs up into small, discrete, and compartmentalized chunks that would lead smoothly to the imminent introduction of two-column proofs. Further attention to language grew during the 1930s, when textbook writers began including vocabulary lists at the end of chapters, as did Welchons and Krickenberger's (1933, 1956).

In addition to retaining all the features of the Wentworth book, it would appear that Smith was also interested in convincing students that a goal of learning geometry was for pleasure. Despite the growing utilitarian tendencies of the geometry curriculum over the ensuing years—all the way to present time—pleasure and aesthetics made occasional appearances in educators' defenses of geometry, though sometimes curtailed by more practical concerns. For example, Reeve wrote, in 1930: "We teachers of mathematics need to train our students to appreciate beauty in form and design" (p. 7). On the same theme, but jumping ahead, Hoffer's 1979 textbook, *Geometry: A Model of the Universe,* writes the following:

> Geometry can be fun. It is satisfying to draw pictures to express ourselves and represent our ideas.... Many people admire geometry for its esthetic and logical aspects.... But geometry is useful in many occupations as well as in daily life. (p. v)

I return to Hoffer's textbook later in the monograph because it also attempted to support the van Hiele theory of the development of geometric reasoning.

In terms of the overall high school sequence, the Committee of Ten upheld the pervasive practice of teaching geometry after one year of algebra. However, it made a rather revolutionary recommendation that demonstrative geometry "should be ... carried on by the side of algebra for the next two years" (p. 113). According to Willoughby (1967), several schools attempted to teach parallel courses in algebra and geometry—following the Committee's recommendations—but considered the experiment a failure. Quast (1968) conjectured that these schools probably continued to teach algebra and geometry as though they were independent subjects. In fact, the report made no recommendations about how teachers were to go about teaching geometry and algebra side by side, and so the result was perhaps not surprising. Even today, especially at the high school level, geometry and algebra continue to be taught as separate courses, though most geometry textbooks require, and assume, algebraic fluency.

A further barrier to implementing the geometry/algebra side-by-side recommendation was established by the Committee on College Entrance Examinations (CCEE), which was organized in 1895 to bring about greater articulation between the colleges and the secondary schools. The CCEE offered a detailed sequence of courses to be used for guiding college admission, which assigned unit values for each subject in each grade, thereby effectively separating the teaching of geometry and algebra. By 1896, the entrance requirements of 432 post-secondary institutions were as follows: 294 required plane geometry, 93 required solid geometry, eight required spherical geometry, and two required conic sections (Quast, 1968).

To this day, the idea that mathematics in the secondary school should be decompartmentalized has never been widely accepted, despite continued, although sporadic, support over the years. The greatest opportunity for decompartmentalization came during the first half of the twentieth century, during which the Laboratory School at the University of Chicago, under the leadership of E. R. Breslich, taught all of its secondary mathematics courses on an integrated basis. However, there were many opponents to this idea. For example, in 1912, Decker wrote:

> So much has been said about teaching [algebra and geometry] side by side—together, if you please—that one is at times tempted to forget that such a combination is, if we may borrow a figure from natural science, a mechanical mixture rather than a chemical compound. Such a method has doubtless its advantages but it has a fundamental disadvantage in obscuring the development of geometry. (pp. 41–42)

The Committee of Ten also recommended that, as long as students have an introductory course in concrete geometry, "both plane and solid geometry can be mastered at this time" (p. 24). Three years later, however, the CCEE was "not prepared to commend [the idea of fusion]" (p. 771)—representing another instance in which the two Committees were less than harmonized. This was expected, given the fundamentally differing aims the two committees had. It should be noted that the question of fusion was contentious amongst mathematicians as well, and for pedagogical reasons. For example, in 1906, Hadamard (1898/1947) wrote:

> Que cette fusion soit preferable au point de vue logique, je le veux bien. Mais il me paraît que, pédagogiquement, nous devons penser tout d'abord à diviser les difficultés. Celle de "voir dans l'espace" en est une sérieuse par elle-même, que je considère pas comme devant être ajoutée tout d'abord aux autres. [I can well accept that this idea of fusion might be preferable from a logical point of view. However, it seems to me that, pedagogically, we should think first of dividing the difficulties. That of "seeing in space" is a serious one on its own, which I do not think should be added to the others (my translation)]. (p. v)

Some later critics of a "fused" approach accused it of providing students with a "hodgepodge of superficialities that tends to general weakening of the subject content."

The Committee of Fifteen: Toward a National Syllabus

In 1912, two decades after the Committee of Ten, the National Committee of Fifteen on Geometry Syllabus, sanctioned by the NEA, published its report, which was much more specific than its predecessor's. Once again, the committee essentially followed the traditional Euclidean approach, but tried to strike a better balance between the rigor of the formalists and the more "utilitarian" approaches. The word "utilitarian," used by Quast (1968), may be somewhat misleading, since it conflated practical approaches (which emphasize applications and aim to serve the everyday needs of students) with intuitive approaches (which *might* include applications but may still aim to serve the cultural or theoretical needs of students). In fact, the committee remained committed to the idea of transfer of training and the development of the faculty of reasoning, as this quotation from the report illustrates:

> a formal treatment of geometry, is necessary purely as a prerequisite to the study of more advanced mathematics, and still more because such treatment has a genuine culture [sic] value, for example, in the assisting to form correct habit in the use of English. (NEA, 1912, p. 12)

This committee did not take into account the fact that transfer of training was not supported by experimental evidence, and so was not compelled to provide alternate goals or methods for the teaching of geometry. It was the members of the International Commission on the Teaching of Mathematics, reporting in 1912, whose main role was to assess what was taking place in American schools, who first recognized the value of these psychological experiments.

The Committee of Fifteen dealt with many of the same issues—on which they had little impact—facing the Committee of Ten, including the inclusion of practical applications, the integration of solid and plane geometry, and the sequencing of geometry and algebra in the high school curriculum. Their allegiance to Euclid was clearly stated:

> With all the experiments at improving Euclid the world has really accomplished very little except as to the phraseology of proportions and proofs; the standard propositions remain, and if geometry had any justifications … most of these propositions will continue to be proved, and should continue to be proved. (pp. 3–4)

This allegiance caused no stir, and not surprisingly: the ICTM had found that the sequence of all the ordinary textbooks used in schools at the time was that of Legendre.

According to Shibli (1932), the primary impact of the Committee of Fifteen, due to its greater specificity, was to promote the movement toward a uniform and nationally consistent syllabus. Additionally, it introduced the idea of different courses according to the needs and abilities of students, a shift that had already taken place in England, as the next section shows. It resolved few other issues save for reducing the number of basal propositions and confirming the idea that some easy propositions should be removed and either postulated or proved informally.

1902: PERRY AND MOORE AND THEORIES OF LEARNING

With the increasing number of students enrolled in secondary school, and the increasing dissatisfaction with the teaching of geometry, John Perry's address to the Section on Educational Science at the 1902 meeting of the British Association for the Advancement of Science struck a deep chord. In it, he noted that

> demonstrative geometry and orthodox mathematics generally are not only destroying what power to think already exists, but are producing a dislike, a hatred for all kinds of computation, and therefore for all scientific study of nature, and are doing incalculable harm. (p. 163)

Perry denounced both the practices of teaching prevalent in England (and the United States) at the time and the aims of teaching geometry assumed by these practices and existing textbooks. He called for a much less theoretical orientation to the teaching of geometry, insisting that experimental geometry and practical mensuration be taught before demonstrative geometry, and that practical applications should be emphasized. Perry also gave increased prominence to the instruments of geometry by suggesting that visual devices, graphical methods, and square paper should be used *throughout* the study of geometry. Most significantly perhaps, he advocated the laboratory method of teaching, one that emphasized both the use of instruments and the use of experimental, intuitive methods, where students would be free to experience and discover on their own.

Perry's recommendations may sound familiar to some issued previously, but there were two underlying differences that made his recommendations more consistent: (1) he called into question the traditional aims of teaching geometry, namely, to teach reasoning and appreciation, and argued for the appropriateness of much more practical aims, given the population of students now in schools; and (2) he did not remain attached to the same traditional Euclidean content that previous policy documents assumed (both in England and the United States) and, instead, envisioned a geometry that was well suited to experimental methods. This consistency may have been at least in part responsible for the deep effect the Perry movement had in England. Wren and McDonough (1934a) noted that by 1932, two courses in mathematics were offered, one styled "pure mathematics" and the other "practical mathematics" (these are roughly equivalent to the later division in courses that led to A- and O-levels).

In the same year, in the United States, mathematician E. H. Moore gave a presidential address to the American Mathematical Society in which he, too, advocated the use of the laboratory method, as well as the more practical side of geometry. In addition, he strongly advocated a more unified program that would blend not only algebra and geometry, but physics too. While Perry was concerned with the teaching of mathematics in general, Moore was especially interested in geometry. But he too, and this is perhaps more significant given that he was a mathematician, suggested radical departures from the traditional Euclidean approach. In fact, he and Perry together veritably led the movement that sought to distance itself from the traditional dependence on Euclid's approach to geometry. As this quotation illustrates, Moore was more interested in the function of geometric axioms in general than in the particular Euclidean axioms—he was perhaps influenced by David Hilbert's systematic study of axioms:

> why should not the student be directed each for himself to set forth a body of geometric fundamental principles on which he would proceed to erect his geometric edifice? This method would be thoroughly practical and at the same time thoroughly scientific. The various students would have different systems of axioms, and the discussions thus arising in the minds of all precisely what are the functions of the axioms in the theory of geometry. (Moore, 1926, p. 48)

In fact, Moore's quotation above presaged some of the beliefs that would later come into play during the "Modern Mathematics" reforms in the 1960s: Moore was interested in the theory of mathematical structures themselves, and in the possibility that students could develop insight into the true foundation of geometry. This made sense historically in that the nineteenth century had given rise to different geometries (such as projective and non-Euclidean) and mathematicians had become quite interested, thanks to Felix Klein's *Erlangen Program*, in the relationships between the different geometries. Given his position on Euclidean geometry, it is not surprising that Moore also advocated the use of more algebra and arithmetic in the teaching of geometry. And as a result, Breslich (1951) pointed out that beginning in 1905, more textbooks were showing the use of algebraic techniques.

For many teachers and textbooks writers, abandoning or loosening the grip of Euclid's approach, however, did not necessarily entail abandoning the traditional aims of geometry instruction. Although early textbook authors were less apt to include such a section on "Why study geometry?" their content and presentation were in line with the discourse of formal argument. Examples include the most widely used textbooks at the time, such as Wentworth and Smith's *Plane and Solid Geometry* (1913).

While the move away from a traditional Euclidean approach caught on, the laboratory method suffered. Teachers frequently cited the expense of equipment as well as the increased time required to prepare laboratory-based classes, concerns that are familiar now, in the context of computer-based technology use. In addition to these constraints, it is highly possible that teachers (as well as textbook writers) had continued to view the instruments of a laboratory as peripheral to geometric understanding, or as dispensable to it—again, an attitude that now characterizes many teachers' views of computer-based technologies. In fact, some went so far as to propose that the use of instruments could only cause distraction. For example, Davison and Richards, two British textbook writers, argued that:

> Too much time spent on experimental and graphical work is wearisome and of little value to intelligent pupils. They can't appreciate the logical training of theoretical geometry, while experiments and measurements of far greater interest can be made in the physical and chemical laboratories. (1907, p. v)

It was perhaps in part due to the stirrings around the laboratory method that Wentworth and Smith's revision of *Plane Geometry* (1913) introduced exercises using the straightedge and compass—in the very first chapter, after only a few definitions. This textbook was used in well over half the secondary schools that taught geometry (Austin, 1919), and adopted for statewide public use by officials in states such as Indiana and Kentucky. The revision also included an increased number of exercises, which were both simpler and more practical.

At the same time, George Myers, a professor of the teaching of mathematics and astronomy at the University of Illinois at Urbana–Champaign, introduced a textbook that combined topics from traditional algebra and geometry as well as basic physical applications. Although Myers's book was well supported by the central states, it was not as popular as the Wentworth and Smith book. However, its influence, based on Moore's vision, can be traced to textbooks later developed at the University of Chicago. In fact, as can be seen in the progressively growing integration of algebraic elements in geometry textbooks over the course of the following half-century, Myers's textbooks helped launch a movement toward the integration of algebra, geometry, and other areas of mathematics. A notable feature of Myers's book was the inclusion of a teacher's manual—a novelty at the time, but a staple of textbook writing now.

The Impact of Other Theories of Learning

Until this time, the theory of "faculty psychology," which rests on the concept of the mind being composed of several faculties, such as reasoning, will, and memorization, had been dominant within the mathematics education community—if only implicitly. Subscribers to the theory felt that a well-trained faculty would be useful in a variety of situations, so that if geometry developed the reasoning faculty, a person would reflect the ability to reason in situations outside of geometry. Psychologist Edward Thorndike showed that, in fact, little such transfer took place, unless "identical elements" were involved across situations. He thus began formulating his theory of connectionism, which grew to be strong between 1920 and 1940 in mathematics education. The basic premise of the connectionist was that ideas (or entire subjects) should be broken up into a series of simple elements arranged in a sequential order, and that each component part should be mastered separately. Thus, the series of mastered elements formed the whole learning situation. Students would master a subject by mastering each of the separate abilities forming the subject, and the challenge for the educator was to identify the best order in which to present the parts.

The application of Thorndike's connectionist theory in school classrooms led to a well-choreographed system of drill, which in later years was called the "show and do" method (Kinsella, 1965). According to Quast (1968), mathematics teachers were quick to adjust their teaching to this method. However, there were many critics who complained that connectionism placed the role of reasoning and understanding in a secondary position. Naturally, given the low emphasis on transfer of training, connectionism was accompanied by more utilitarian goals, which encouraged the idea of teaching what would be immediately useful. The emergence of the Gestalt school of psychology, at approximately the same time, provided an antidote to Thorndike's connectionism. Gestalt theory was more concerned with the *whole* task, emphasizing insight, organization of material, and the search for patterns. Gestalt theory proposed that what an individual perceived as the result of sensory stimuli depended not only on the setting or "ground" associated with the stimuli, but also on the store of meanings, understanding, and attitudes the perceiver brought to the experience. Educators certainly took note of this theory: for example, the idea that the whole concept of geometry as a logical system was more significant than the learning of a particular theorem or concept emerged in the 1930s. And according to Fehr (1953), after 1940, features from both theories of learning (connectionism and Gestalt) were used.

One can hardly claim, however, that faculty psychology went away. Even in 1965, Kinsella was reiterating findings established decades before: "The amount of transfer from the training experience to the new situation increased with the number of common components possessed by the two, and with the extent to which the learner had recognized the common elements" (p. 75). Evidently, Smith's (1928) plea on the power of geometry to develop logical thinking was shared by many geometry teachers, and was much more than a statement about psychological theory: it was intimately related with a specific view of mathematics, of geometry, and of the goals and methods of teaching geometry.

> If however, the knowledge of how to arrange a logical proof in geometry can be made of no value to us in other fields in which deductive logic can be applied; if the perfection of geometry does not give us an ideal of perfection that helps us elsewhere in our intellectual life; if the succinctness of statement of a geometry proof does not set a norm for statements in non-mathematical lines; if the contact with absolute truth does not have its influence upon the souls of us; if the very style of reasoning does not transfer so as to help the jurist, the physician, the salesman, the publicist, and the educator; if the habit of rigorous thinking, which is usually first begun in demonstrative geometry, is not a valuable habit elsewhere, if a love for beauty cannot be cultivated in geometry so as to carry over to stimulate a love for beauty in architecture—then let us drop demonstrative geometry from our required courses. (p. 14)

One can detect, in this quagmire of psychological theories, underlying philosophical differences about the aims of education, differences that persist today, and that would later be compounded by the increased dominance of progressivist ideas first championed by Herbert Spencer and propagated by Dewey and Piaget (see Egan, 2002).

1920s AND 1930s: COMMITTEES AND THEIR REPORTS REDUX

In 1923, the National Committee on Mathematics Requirements (NCMR), organized under the auspices of the Mathematical Association of America, published an influential report. The purpose of the NCMR was to give "national expression to the movement for reform in the teaching of mathematics ... which lacked the power that coordination and united effort alone could give" (p. vi). And unlike the committees previously discussed, this one contained many notable mathematics educators and thus relatively fewer mathematicians. The influence of the NCMR's report was apparent in the textbooks in the years that followed—most cited it in their prefaces. By this time, it could be assumed that most students would have had some geometry in the junior high school. This permitted the senior high schools to treat mathematics courses as electives. However, the NCMR (1923) report advised that: "every standard high school should not merely offer courses in mathematics ... but should encourage a large proportion of students to take them" (p. 27).

At the high school level, the NCMR report was quite traditional, at least from a content point of view. The principal purposes of teaching plane demonstrative geometry were

> (1) to exercise further the spatial imagination of the students, (2) to make him familiar with the great basal propositions and their applications, (3) to develop understanding and appreciation of a deductive proof and the ability to use this method of reasoning where it is applicable, and (4) to form habits of precise and succinct statement, of the logical organization of ideas and of logical memory. (p. 34)

These purposes were set within broader aims of mathematics articulated in the report: the practical aim was concerned with the ability to use arithmetic, algebra, and geometry with understanding in everyday life; the disciplinary aim was concerned with developing correct habits and attitudes; and, the cultural aim was concerned with the appreciation of geometric form, logical reasoning, and the "power of thought, the magic of the mind" (p. 10). It is interesting to note that geometry seemed to be the prevalent vehicle for the cultural aim, even though the NCMR singled it out as most suitable for achieving the practical and disciplinary aims. While the report retained some of the spirit of "transfer," it noted that the disciplinary value of geometry rested on the manner in which it was taught, thereby transferring (!) much of the responsibility of achieving "transfer" onto the teacher and, to a lesser extent, the textbook. This did not stop textbooks from aspiring for transfer, as Clark and Otis's (1925, 1927) preface showed: "we teach geometry primarily for the purpose of training the student in the methods and habits of thought that result in power to reason and analyze, to discover, and to prove in a logical manner that which has been discovered" (p. iii).

The Use of Motion and Rigid Motion in Geometry

In terms of content, the NCMR made one recommendation that distinguished it from previous reports. It advocated the use of rigid motion—namely, transformations—for "gaining greater insight and saving time" (p. 30), citing the success of the French in doing so. In fact, several decades later, other Europeans would take up transformations wholeheartedly: In 1967, Dienes and Golding would teach transformations and groups to elementary school students in Hungary and Australia; and, in 1964, Georges Papy would develop in Belgium an entire school curriculum with heavy emphasis on group properties and vectors.

The use of motion in geometry already had a long and troubled history. In one of Zeno's paradoxes, known as the tortoise and Achilles, the tortoise argues that no matter what kind of head start Achilles grants him, he will surely win the race. Just as Achilles makes up the distance between them, the tortoise has, in the meantime, traveled some distance of his own. The paradoxes warned the ancient Greeks of the difficulties of talking of motion. While the paradox of motion here arises from the inability to conceive that a sum of an infinite number of distances could be a finite distance, not from any problem of superposition, it may well have contributed to the general discomfort with the use of motion apparent in Euclid's *Elements*. Yet, despite Euclid's best attempts, a certain type of motion snuck its way into the axiom of superposition, in which two shapes can be shown to be congruent by placing one on top of the other.

Superposition, which is used to prove Euclid's first triangle congruence theorem (Proposition 4), assumes that a figure can be moved in space without changing its size or shape. Mathematician Giuseppe Veronese voiced this discontent with Euclid's method:

> Since geometry is concerned with empty space, which is immovable, it would be at least strange if it was necessary to have recourse to the real motion of bodies for a definition, and for the proof of the properties of immovable space. (quoted in Heath's translation of Euclid's *Elements*, 1926, pp. 226–227)

Others expressed concern about the uses of superposition in three dimensions, where it was more difficult to imagine, as in superimposing one sphere on another.

However, it proved difficult to keep the language of motion out of geometry altogether. After all, motion had a rich history in mathematics, despite Euclid. For example, Clairaut (1741) used motion to show that the sum of the angles of a triangle was constant: "let's examine what would happen to this angle [C] if lines *AC*, *BC* came together or moved away from each other. Suppose for example that *BC*, which turns around point *B*, moves away from *AB* and approaches *BE*; it is clear that while *BC* turns, angle *B* would open continuously and that, on the contrary, angle *C* would get smaller and smaller" (p. 63).

Unrest over the use of motion persisted in North American and British schools during the 1920s and 1930s. The importance of the place of motion in geometry was well illustrated in the publication of the British Mathematical Association in 1923, in which motion was extensively discussed as the second of the most "disputed points" in geometry teaching. One has only to compare the modern student dragging points around in a dynamic geometry environment with Bertrand Russell's (1903) quotation below to see the changes in thinking about motion that have occurred during the past century:

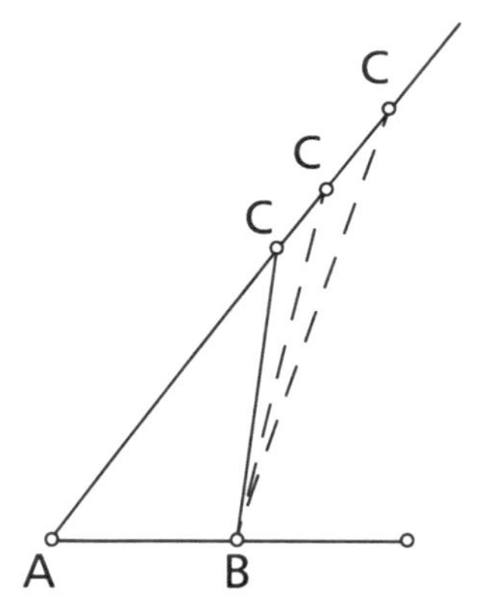

Figure 3.5. Clairaut's 1741 use of motion.

> To speak of motion implies that our triangles are not spatial, but material. For a point of space *is* a position, and can no more change its position than a leopard can change its spots. The motion of a point of space is a phantom directly contrary to the law of identity: is it the supposition that a given point can be now one point and now another? (p. 405)

While motion that actually required movement proved difficult to ban, Euclid made no formal use of rigid motion, or transformations, in the *Elements*—this, despite Appollonius's earlier use of them (Coolidge, 1963). The two types of motions, though presenting very different mathematical and philosophical problems, were not easily separated. Bertrand Russell felt obliged to clarify the distinction as such: "The fact is that motion, as the word is used by geometers, has a meaning entirely different from that which it has in daily life.... Motion is a certain class of one–one relations" (quoted in Mathematical Association, 1929, p. 33). For the purposes of this monograph, I distinguish *rigid motion* from *motion*, and use *rigid motion* to describe transformations (translations, reflections, glide reflections, and rotations), which do not actually require physical movement, and *motion* to describe the more daily life interpretation in which mathematical objects actually travel to a new position. To put it more clearly: a 1-1 distance-preserving transformation can be described purely by a correspondence of points, and does not require actual movement of the mathematical objects, but such a transformation can also be visualized as being carried out by a physical displacement that moves all points of the object to their new positions defined by the correspondence.[7] The introduction of transformation geometry and the definition of congruence through the use of transformations formalized Euclid's conception of superposition.

Eventually, rigid motion did make a more explicit appearance in geometry, primarily through its use in the constructions of elementary geometry. At the time of Legendre's reworking of Euclid, in 1794, very little had taken place. In fact, it was not until the late nineteenth century that any significant work was done. At this time, Felix Klein discovered a new way of characterizing all possible geometries that was based on transformations (though not just distance- or similarity-preserving ones that are found in school textbooks today). The impact of his work—both mathematical, and, later, pedagogical—was significant: transformations became a fundamental concept in research geometry, and were eventually widely used in European schools. The late development of transformations may account for the fact that until the mid 1960s, in the United States at least, studying geometric transformations (rigid motions) was deemed too hard for high school students.

The NCMR was motivated to recommend the use of rigid motion in order to alleviate the superposition problems found in Euclid, and to help develop the functional concept for students. Its report signaled the first policy-driven attempt in the United States to include transformations in the geometry curriculum. And while several textbooks included transformations as topics additional to traditional geometric approaches (see Hill, 1887, and Young and Schwartz, 1915, for early instances), it would not be until several decades later, in the late 1960s, that a truly transformational approach to geometry would form the basis of a high school textbook (see Coxford & Usiskin, 1971). In 1933, Betz bemoaned the lack of attention to transformations on both mathematical and pedagogical grounds, and blamed the conservative nature of high school geometry:

> In geometry there has been a growing conviction that the use of symmetry as a fundamental principle, through the idea of rotation and folding, is the best device we have for making many geometric properties clear intuitively. Unfortunately, our conservative high school geometry has avoided the extensive employment of symmetry as a tool, although a thoroughly rigorous treatment is possible on this basis, and although great mathematicians have long given enthusiastic support to this mode of approach. (p. 110)

Not long after the NCMR report, the committee of the NCTM on geometry published its reports in *The Mathematics Teacher* in 1931, 1933, and 1935. The committee made 47 suggestions and recommendations, based on a review of all the literature dealing with the teaching of geometry published since 1900. Although the committee's reports were very comprehensive, covering the entire span of school geometry both in the United States and in Europe, they had negligible impact on the shape of the curriculum. One of the committee's recommendations stemmed from the French approaches to geometry based on transformational geometry. They noted that Meray, for example, had written an elementary geometry text in 1874, making extensive use of symmetry, translations, rotations, and reflections, and that several others had followed suit, including Hadamard (1898/1947), André and Lormeau (1908), and Bourlet (1928). However, as mentioned above, it would take another 30 years before American textbooks would follow this particular recommendation. Usiskin (1969) placed partial blame on this slow adoption to the confusion between "rigid motion," as it applies to transformations, and the imaginary "motion" used in superposition. Those who had criticized the use of superposition as an invalid method of proof extended their discontent with motion to the use of transformations.

The Changes, and Lack of Changes, in the Geometry Curriculum

Although the NCMR's recommendation on transformations was not heeded, the report did have one immediate impact.[8] With some concern, the NCMR noted that the college entrance examinations were exerting undue pressure on the high schools by overly influencing the curriculum. The CEEB responded by reducing the number of "book theorems" students would be responsible for in plane geometry to 89, and increasing dependence on "originals."

In terms of the junior high school curriculum, the NCMR may have had little influence. A 1927 survey of 27 junior high textbooks showed strong parallelism in the topics suggested, but found that less than 50% of the subject matter was used by all authors (Wren & McDonough, 1934a). Judging by the number of pages taken up by each topic, arithmetic and algebra were found to be the most important topics. The lesser focus on geometry may have been due to the enduring lack of precise aims of intuitive geometry. Quast (1968) suggested that the junior high school mathematics courses were also becoming increasingly "general," in that the different topics were becoming less differentiated. As McCormick (1929) noted, though, "there is no clear or definite agreement among mathematicians and general educators as to what consists general mathematics" (p. 162). Once traditional boundaries were blurred, the notion of general mathematics tended increasingly to arithmetic and algebra.

In assessing the progress of mathematics teaching related to geometry during the first quarter of the twentieth century, David Eugene Smith (1926) cited various practices that he saw as widespread during this time. The first was the growth of intuitive geometry in the elementary school.

The second was a reduction in the number of theorems presented along with an increase in the number of original exercises, which had generally decreased in difficulty. Christofferson (1938) analyzed trends in textbooks between 1843 and 1931, and found a decrease in the number of propositions given. Freeman (1932) reviewed textbooks from 1896 to 1932 and found close agreement in terms of the order of introduction in larger divisions of content, but variation in the sequence of propositions in current texts. In contrast, he found that later books presented construction theorems following the theorems on which they depended instead of leaving them to the end of the chapter, as was done in earlier books. Constructions and exercises were more prevalent in later books as well. Additionally, increased emphasis was placed on geometric instruments and their use, on the values to be gained by studying geometry, and on historical connections and background. Finally, the use of symbols and abbreviations had increased, as had the inclusion of tables, illustrations, and varied typography. Freeman also noted the presence of simpler language and less formal style. Wren and McDonough (1934a, 1934b) summed up Freeman's findings as follows: "Contrary to a belief frequently held, it is found that textbooks in plane geometry have experienced notable changes since 1900 ... all aim toward assisting the pupil to independent discovery of truths and their proofs" (pp. 293–294).

The changes documented by Freeman were embodied in Wells and Hart's widely used *Progressive Plane Geometry* (1935), which was in many ways very similar to Wentworth's *Plane Geometry* (1892) as well as Wentworth and Smith's *Wentworth's Plane Geometry* (1910). The textbook *Progressive Plane Geometry* begins with a four-page history of geometry, and includes a section on "why you should study geometry": first, "for the fun of it... for the fun of using your brains just as you get fun from using your muscles"; second, "for the possible usefulness of it," whether that be an appreciation of geometry in the physical world or applicability in a trade or vocation; third, for the indirect benefit of "the training in logical reasoning, argumentation and precision in language such study provides" (p. 5). The third reason suggests transfer of training, which Wells and Hart supported explicitly in their approach to proofs (Donogue, 2003). In addition, several features were included, ostensibly to increase the accessibility of geometry for students, including two-column proofs, less formal language and explanations; some initial "experimental geometry" (drawing figures, measuring line segments and angles, and making observations in response to queries on, for example, vertical or supplementary angles); and, a section on "how such proofs are discovered," which modeled an interior dialogue of questions and answers that would lead to an acceptable proof.

But there were even more radical departures made in textbooks at the time. Swenson, the author of *Integrated Mathematics with Special Applications to Geometry* (1935), wrote that some progress had been made to reform the teaching of algebra, "but geometry has remained fundamentally the *static* geometry of Euclid" (preface, n.p.). The goal of his textbook was to make "mathematics of the tenth year more comprehensive, dynamic, and functional in character" (preface, n.p.). To this end, the first chapter illustrated the role that motion could play in geometry—and more in terms of actually moving objects than in terms of transformations. For example, lines were linked to moving points: "If a point moves it generates or describes a *line*. If the point keeps moving in the same direction we obtain a straight line...' if the point changes the direction of motion at certain intervals we have a *broken line*...; if the direction of the moving point is changed continuously we have a *curved line*" (p. 5). In Chapter 7 of his textbook, Swenson again used ideas of motion to explore the geometry of space. Solids were viewed as being generated by the movement of planar areas through space: "A solid generated by a polygon moving parallel to its original position in such a way that the vertices of the polygon move on parallel straight lines not in the plane of the polygon is called a *prism*..." (p. 259). The emphasis on generating two-dimensional lines from one-dimensional points in motion and three-dimensional solids from two-dimensional figures in motion even permitted a brief introduction to four-dimensional entities. Paper-folding exercises introduced line and point symmetry. Swenson can be seen as promoting "functional thinking"—assessing the dependence of one variable on another—in the classroom, something that would gain much popularity in later algebra courses.

As with Swenson's textbook, Blackhurst's *Humanized Geometry* (1935) also had some more radical elements to it, including an extensive coordination of inductive and deductive approaches.[9] Most other textbooks at the time began with some inductive approaches but then generally tried to illustrate the drawbacks of the approach in order to motivate the (exclusive) use of deductive methods. Even today, the split between inductive and deductive thinking persists in many textbooks. It appears that, at least in 1935, the split was insurmountable: Beatley (1935) published a report, that included results from a questionnaire given to 101 teachers in eight states, showing that teachers unequivocally wanted to maintain a distinction between informal and deductive geometry. They viewed the former as being about facts and the latter about logical thinking. Geometry was still seen as *the* primary locus of deductive reasoning in the school mathematics curriculum, and mixing methods could only detract from the objectives of geometry education. This, despite the Committee on Geometry's findings that few teachers, in practice, called attention to the logical chains of theorems, to the gaps in Euclid's logic, to the appreciation for the nature of a mathematical system, and to the need for undefined terms and arbitrary assumptions.

Another important book contemporary to the previous two was Fawcett's *The Nature of Proof* (1938), which was carefully researched. Kinsella (1965) called it one of the most serious attempts to implement disciplinary and transfer aims in the field of geometry. *The Nature of Proof* included many instances of analyzing problems from school life, home situations, politics, and advertisements. The primary goal was to develop students' understanding of a logical system and the nature of proof, but it related the characteristics of deductive thinking to real-life situations. A standardized, statewide geometry test was given. The class median for pupils in the experimental group (using Fawcett's book) was far above the state median. This result was especially significant because the greater emphasis on logic in nonmathematical situations meant that less time was spent on geometric concepts and facts. Alas, other forces usurped the promising research. Among them: the New York syllabus only allotted 5% of its items to reasoning in nonmathematical situations, making Fawcett's textbook unusable to teachers in that state.

The last practice that Smith saw as being increasingly widespread was the strong movement to cover the essential parts of plane and solid geometry in one year. Smith saw all these changes as improvements and was heartened to observe that it was indeed possible to make changes in the hitherto hopelessly static subject of geometry. Even if Smith saw these improvements as unquestionable, their causes were more difficult to discern. Had the various committees exerted influence? Betz (1936) argued that they had not, claiming they had all been ineffective by failing to include "professional students" of curriculum-making, educational psychology and sociology, as members. Betz believed that the omission prevented committees from appreciating the curriculum as a whole.

With the depression in the 1930s, increased emphasis was placed on the practical aims of geometry education. As Kinsella (1965) noted, "social utility was a major factor in determining what was taught" (p. 11). In this period of time, geometry became especially difficult to defend, and educators responded by pointing—once again—to the benefits of training in deductive thinking. Betz (1930) wrote that:

> demonstrative geometry offers the simplest and most convenient introduction to postulational thinking which has yet been devised. Hence, it may be claimed that the teaching of demonstrative geometry is not only justifiable, but absolutely essential, because of its permanent devotion, in singularly pure and significant form, to the one procedure that promises valid conclusions on the basis of clearly formulated assumptions. (p. 151)

The problems that students were encountering in learning this kind of postulational thinking began to be attributed to teachers instead of to the subject itself. Reeve (1936) went so far as to criticize the "stupid way in which many teachers continue to present both algebra and geometry" (p. 20).

Even before the depression, however, the status of geometry had declined. By 1928, there was a significant drop in the percentage of students taking geometry: whereas 30.87% of students took geometry in 1910, only about 20% did by 1928 (Breslich, 1933). This drop may have resulted from the movement to make mathematics an elective subject in high school (by 1936, 17 states had eliminated mathematics from their lists of prescribed courses). Quast (1968) attributed the 10% drop to a decline in the popularity of geometry, but the increase in educational testing during this period may also have contributed to the decline. The depression itself, motivating students to take subjects that could be helpful immediately in gaining employment, was surely a major force at work in the decline of high school mathematics in the 1930s. However, there were probably larger forces at work too, more related to the developments in mathematics as a whole than to any factors within schools, as the next section argues.

1935: THE BOURBAKI GROUP AND ITS IMPACT ON GEOMETRY

In 1900, David Hilbert delivered a famous lecture in which he listed 23 important, unsolved problems. These problems were to have a strong influence on twentieth-century mathematics, in the sense that many mathematicians devoted their energies to solving them. Only three out of 23 were in geometry. Much later, in 1976, a symposium was held on the mathematics arising from Hilbert's problems, and a group of mathematicians proposed 28 important, unsolved problems, none of which included geometry (although there were problems in more current derivatives of geometry such as algebraic geometry, differential geometry, and geometric topology). So, one can conclude that by 1976, geometry was no longer seen as an important area of research in mathematics.

Whiteley (1999) pointed out that a field of mathematics "dies" when it is no longer viewed as an "important" area of mathematical research, and argued that geometry "died" in this sense through the 1920s–1940s, at least in North America and parts of Europe (but it did survive in pockets, in countries such as Hungary, Germany, Switzerland, Austria, and Russia). Geometry in the education system then followed in a predictable decline, starting at the graduate-school level, then continuing on down into the high school and elementary classrooms. In the Canadian context, when H.S.M. Coxeter, one of the few remaining great geometers, retired at the University of Toronto many years ago, the mathematics department decided to give up the position in geometry, and opted instead for new hires in "hotter" areas of mathematics.

Whiteley described the predictability of the decline: as research in geometry declined, the importance of teaching geometry in graduate programs also declined, as did the number of faculty offering courses in geometry. Increasingly, graduate programs contained no researchers in geometry. This state of affairs gave rise to a generation of people responsible for teaching undergraduate mathematics "who have not experienced geometry as an important, lively field of current mathematics, and who may not have studied any geometry during their graduate studies" (p. 8). As a result, logicians and historians of mathematics started taking responsibility for teaching geometry to undergraduates. As E.T. Bell (1940) noted: "The geometers of the twentieth century have long since piously removed all these treasures to the museum of geometry where the dust of history quickly dimmed their luster" (p. 323). Indeed, Whiteley observed that both of these groups will often teach geometry as an important *past* accomplishment (and as an exercise in logical proofs) but not as a continuing source of new mathematics: "many undergraduate geometry courses wear a veneer of geometric language without any playful geometric and visual spirit in the problems, the solutions, or the presentation" (1999, p. 8).

Over time, the decline of geometry in the universities reached the point where algebra and analysis became the core areas in the undergraduate curriculum. Only 40% of universities required plane geometry of their college students. By the 1960s, geometry was largely relegated to being a service course for future high school teachers. The marginalized place of geometry in the curriculum was well reflected by the development of the AP program in 1955. Initially sponsored by the Ford Foundation, and designed to develop rigorous, college-level course curricula and assessments for high school students, the AP program has never, to this day, included an advanced placement geometry test. The incipient "new math" movement effectively pushed analytic geometry—which used to be a first-year course—down into high school. The race to calculus had begun! The shifting of college subjects also had an effect on the topics included in high school geometry. Without analytic geometry, for example, there was little reason to treat loci.[10] And the only reason to do solid geometry was to derive volume theorems in calculus, for which the deductive apparatus behind Euclidean solid geometry was seen to be overkill.

After a few more decades, the decline that started in research geometry led to a generation of high school teachers who either had no undergraduate courses in geometry, or only had a "geometry for teachers" course in college. The lack of availability of geometry courses implicitly communicated to intending teachers that geometry is not a central part of modern mathematics. The final stage of the decline produced a group of teachers who were uncomfortable with open-ended problems in geometry and who viewed geometry as being much less important to their students than core areas like functions, algebra, and calculus. From there, curriculum writers, the textbook writers, and even parents developed a sense that geometry was an optional topic (see Pedoe, 1998). (Of course, the decline in geometry described by Whiteley did not preclude the continued development of geometry programs and textbooks at the school level. K–12 students continued to take geometry; however, Whiteley's point is that geometry ceased being represented to prospective teachers [and, eventually, to their students] as important and alive.)

In terms of twentieth-century views, one can even discern a broad cultural perception that geometry is marginal within mathematics. People often associate mathematics with numbers, formulae, algebra, and maybe analysis. Butterworth's (1999) book, titled *What Counts: How Every Brain is Hardwired for Math*, virtually equates mathematics with numbers and the abilities based on them (e.g., algebra). Similarly, as Whiteley (1999) notes, Howard Gardner equates "mathematical intelligence" with a single approach involving logical sequences of formulae and sentences, separating it from both "visual intelligence" and "kinetic intelligence." Even mathematicians portray mathematics—at least higher mathematics—as being essentially about logical intelligence, downplaying or ignoring the role of the spatial and visual intelligence that is so important in geometry. In fact, when "geometry" comes up, Whiteley argues that no time is lost in translating it into analytic terms, and treating the geometry as a means to illustrate the "important areas" of mathematics such as algebra and calculus.

The decline of geometry at the research level in undergraduate and graduate mathematics can be attributed to several factors, including the growth of new fields in mathematics, the paucity of geometry problems in Hilbert's set of unsolved problems, and perhaps even geometry's own long history. However, the advent of the Bourbaki Group in 1935 was likely a strong factor as well. Nicolas Bourbaki was the pseudonym used by a group of mainly French mathematicians who wrote a series of books on modern advanced mathematics. The group's ultimate goal was to base all of mathematics on set theory, and their method was one of utmost rigor and generality. They identified mathematics with language and formulae, and had a very strong influence on the style of mathematics—an influence that can be felt to this day. For example, one of the members, Jean Dieudonné, urged a "strict adherence to the axiomatic methods, with no appeal to the "geometric intuition," at least in formal proofs: a necessity which we have emphasized by deliberately abstaining from introducing any diagram in the book" (quoted in Brown, 1999, pp. 173–174).

The disdain of images ran strong among the Bourbaki, as the famous story of Claude Chevalley, another member, illustrates. Chevalley is said to have been giving a very abstract and algebraic lecture to a group of students at Berkeley when he got stuck. After a moment of pondering, he turned to the blackboard, and, trying to hide what he was doing, drew a little diagram, looked at it for a moment, then quickly erased it, and turned back to the audience and proceeded with the lecture. Eisenstein (1983) helped reveal the loss of understanding that the absence of images can give rise to when he reports that in the eleventh century, because no diagrams had come down to them, "the most learned men in Christendom engaged in a fruitless search to discover what Euclid meant when referring to interior angles" (p. 296).

While disdained by the Bourbaki, visual arguments were used extensively by the ancient Greek traditions. One has only to consider the fact that the original meaning of the Greek word δε'ιχνυμι (deiknumi), "to prove," was to make visible or to show. In addition, Euclid's own work had already several times been visually "enhanced." For example, Dee's version of Euclid's *Elements*, published in 1570, included pop-up figures in Book IX, together with clear instructions for their use (see Figure 3.6).

And there was also Byrne's version of Euclid, published in 1847, in which colored diagrams and symbols are used instead of letters—this supposedly for "the greater ease of learners" (as quoted in the title) so that, for example, an angle would be referred to by a color, not a name. Figure 3.7 shows Byrne's version of Proposition 47, the Pythagorean theorem. Byrne claimed that students could "learn geometry in less than a third of the time necessary for traditional methods" (p. xii). Klotz (1991) ascribed the disappearance of Byrne's Euclid to the high costs associated with the kinds of new technologies needed to try new forms of visualization. However, Byrne's Euclid would have had a difficult time surviving the Bourbaki wave of influence.

The decline of geometry in research mathematics, and then in undergraduate programs, added extra confusion to the existing muddle of the aims of geometry education. The aim of teaching geometry that was connected to its importance in mathematics, which was just emerging, slowly lost sway. And, as we have seen, the aims connected to both "transfer of training" and "faculty psychology" were also receding. In fact, one could argue that clarity in the aims of teaching geometry would not emerge again until the early 1990s, when new tools and applications emerged on the scene, and a renewed appreciation of the role of visualization in

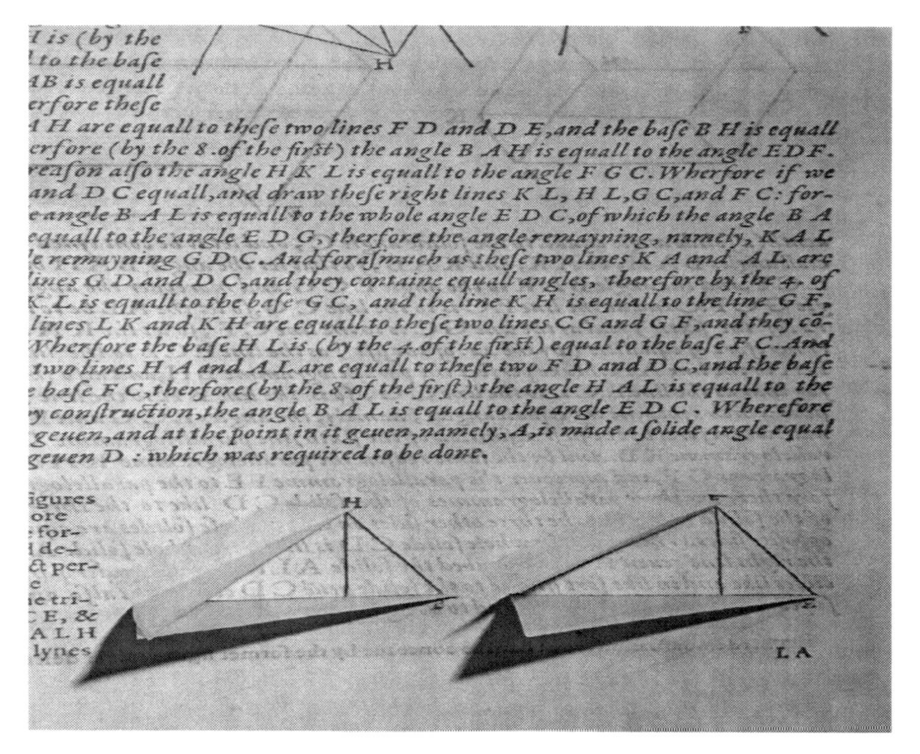
is (by the
to the baſe
AB is equall
erfore theſe
H are equall to theſe two lines F D and D E, and the baſe B H is equall
erfore (by the 8. of the firſt) the angle B A H is equall to the angle E D F.
reaſon alſo the angle H K L is equall to the angle F G C. Wherfore if we
and D C equall, and draw theſe right lines K L, H L, G C, and F C: for-
e angle B A L is equall to the whole angle E D C, of which the angle B A
equall to the angle E D G, therfore the angle remayning, namely, K A L
e remayning G D C. And foraſmuch as theſe two lines K A and A L are
ines G D and D C, and they containe equall angles, therefore by the 4. of
K L is equall to the baſe G C, and the line K H is equall to the line G F,
lines L K and K H are equall to theſe two lines C G and G F, and they cō-
Wherfore the baſe H L is (by the 4. of the firſt) equal to the baſe F C. And
two lines H A and A L are equall to theſe two F D and D C, and the baſe
e baſe F C, therfore (by the 8. of the firſt) the angle H A L is equall to the
y conſtruction, the angle B A L is equall to the angle E D C. Wherefore
geuen, and at the point in it geuen, namely, A, is made a ſolide angle equal
geuen D: which was required to be done.

igures
ore
for-
de-
& per-
e
ie tri-
E, &
A L H
lynes

Figure 3.6. Pop-up figures in Dee's version of Euclid's *Elements*.

mathematics developed. But before then, many events, unconnected to developments in mathematics, continued to influence the teaching and learning of geometry in North America.

One was the publication in 1940 of Birkhoff and Beatley's *Basic Geometry*, which was not as influential at the time of its publication as it would later become. This textbook represented the most significant attempt to modify the postulational structure of the traditional Euclidean approach. Among the many improvements[11] designed specifically for high school students was the use of a set of four postulates. Table 3.2 shows both Euclid's and Birkhoff's postulates: the differences are quite striking, and illustrate Birkhoff's debt to Legendre's introduction of numerical quantities. The new approach was defended as follows: "Taking for granted those fundamental properties of number also leads to many other simplifications and gives us a tool of the greatest power and significance" (p. 4). So, instead of following the trend of broadening the postulational base to include obvious propositions, this approached replaced the traditional Euclidean-based set of postulates with an equivalent set that permitted

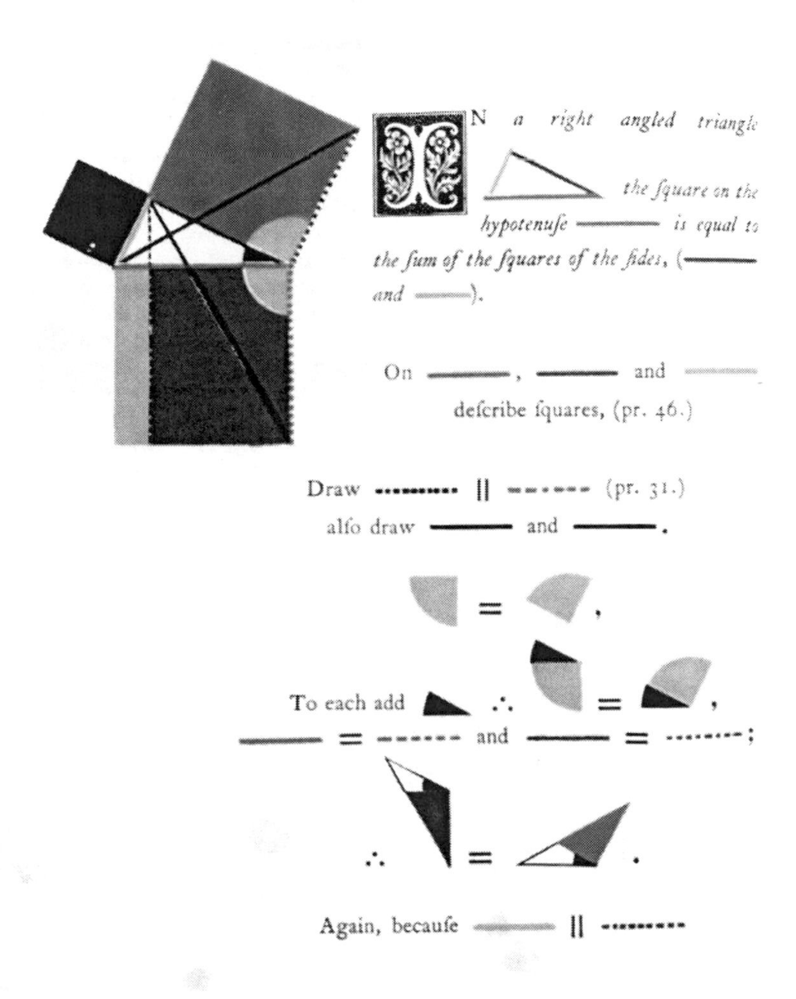

48 *BOOK I. PROP. XLVII. THEOR.*

IN *a right angled triangle* *the ſquare on the hypotenuſe* is *equal to the ſum of the ſquares of the ſides*, (and).

On , and deſcribe ſquares, (pr. 46.)

Draw || (pr. 31.)
alſo draw and .

= ,

To each add ∴ = ,

= and = ;

∴ = .

Again, becauſe ||

Figure 3.7. A page from Byrne's version of Euclid's *Elements*.

greater flexibility. For example, one postulate assumed the side–angle-side similarity proposition, from which side–angle–side congruence is a corollary, thus avoiding the need for superposition (Euclid's method of determining whether two triangles were congruent). In addition, a strong emphasis was placed on understanding what postulates mean and how

Table 3.2. Comparison of Euclid's and Birkhoff's postulates

Euclid's Five Postulates	*Birkhoff's Four Postulates*
1. A straight line segment can be drawn joining any two points. 2. Any straight line segment can be extended indefinitely in a straight line. 3. Given any straight line segment, a circle can be drawn having the segment as radius and one endpoint as center. 4. All right angles are congruent. 5. If two lines are drawn that intersect a third in such a way that the sum of the inner angles on one side is less than two right angles, then the two lines inevitably must intersect each other on that side if extended far enough.	1. The points A, B, ..., of any line can be put into 1:1 correspondence with the real numbers x so that $\|xb - xa\| = d(A,B)$ for all points A and B. 2. One and only one line, l, contains any two distinct points P and Q. 3. The half-lines (or rays) $l, m, n, \ldots$, through any point O can be put into 1:1 correspondence with the real numbers a (mod 2) so that if A and B are points (other than O) of l and m, respectively, the difference $am - al$ (mod 2) of the numbers associated with lines l and m is $m\angle AOB$. 4. If in two triangles ABC and $A'B'C'$ and for some constant $k > 0$, $d(A', B') = kd(A, B)$, $d(A', C') = kd(A, C)$, and $m\angle B'A'C' = m\angle BAC$, then also $d(B', C') = kd(B, C)$, $\angle C'B'A' = \pm m\angle CBA$, and $m\angle A'C'B' = \pm m\angle ACB$.

they affect mathematical thinking. The need for undefined terms was also clearly explained. And frequent references were made to three-dimensional geometry and some to modern geometries. For example, after the introduction of fundamental principle 5 on similarity, in Chapter 2, Birkhoff and Beatley (1940) included the following exercise: "Draw a triangle on any convenient sphere (a tennis ball will do). Then extend two of the sides until each of these sides is double its original length.... What can you say about the third side and the other two angles? Do this for several triangles" (p. 61).

Birkhoff and Beatley especially emphasized the importance of making the importance of mathematical communication, as the following quotation shows:

> student articulate about the sort of thing that hitherto he has been doing quite unconsciously. We wish to make him critical of his own, and others', reasoning. Then we would have him turn this training to account in situations quite apart from geometry. We want him to see the need for assumptions, definitions, and undefined terms behind every body of logic; to distinguish between good and bad arguments; to see and state relations correctly and draw proper conclusions from them. (p. 5)

The authors also made a break from the transfer-of-training tradition most completely articulated by Fawcett's textbook (see above). Instead of making the transfer easier by providing explicit connections between mathematics and everyday situations, Birkhoff and Beatley went the other way, toward a mathematical discourse and away from the utilitarian one:

> In order to meet the demands of our daily lives, we must know how to argue and how to prove things.... we must be able to distinguish between good and bad reasoning.... What we need is a series of abstract and quite impersonal situations to argue about in which one side is surely right and the other surely wrong. (p. 12)

However, they did not relinquish their attachment to transfer:

> It would be difficult to prove that the study of the subject necessarily leads in any large measure to those habits, attitudes, and appreciations which its proponents so eagerly claim for it. But it would be even more difficult to prove that other subjects of instruction can yield these outcomes as easily and as surely as can demonstrative geometry in the hands of an able and purposeful teacher. (p. 6)

While their textbook was well received in mathematics education circles, it did not become popular. Perhaps it was too radically different to be accepted by teachers. Despite this initial poor reception, it would later influence[12] the books by the School Mathematics Study Group (SMSG), one of the major reforms of the 1960s.

1944: POST-WAR DEVELOPMENTS

Reporting in 1940, the Committee on the Function of Mathematics in General Education of the Progressive Education Association stated that the general aims of mathematics were to meet the needs of students and develop personal characteristics essential to democratic living. These practical aims hardly cried out for increased attention to mathematics. However, only 2 years later, with the entry of the United States into World War II, and the accompanying inadequacies revealed in mathematics programs through the deficiencies shown by inductees, many people began to pay more attention to mathematics. Many calls were made for revitalizing mathematics programs, and these were accompanied with much more traditional aims than the ones professed just 2 years earlier. Included in these traditional aims was the notion of geometry being taught for theoretical rather than practical purposes. In fact, Mallory and Fehr (1942) had decried the prevalence of practical aims that had steadily been encroaching the geometry curriculum:

> one movement that has weakened the mathematics program is the introduction of an excessive amount of Reasoning in Life Situations into the subject of geometry. In many cases this has resulted in befuddled thinking and a lack of knowledge of plane geometry. (p. 292)

The tug-of-war between the two sides of the aims issue produced few, but somewhat conflicting changes in the geometry texts. Hlavaty's (1950) study of textbooks from 1930 to 1948 found that one side managed to gain by reducing the number of propositions and increasing the amount of "Reasoning in Life Situations" while the other side gained by increasing the number of "originals" and introducing materials from analytic geometry. Once again, the resolution of discrepancies over aims and content was mostly resolved by the addition of materials to an increasingly crowded curriculum. However, the basic content was quite homogeneous, consisting of a review of junior high school geometry, perpendicular and parallel lines, propositions of quadrilaterals, congruence of triangles, inequalities in triangles and circles, propositions of line segments and angles in circles, angles and area of polygons, propositions of similar polygons, propositions of regular polygons, and measurement of the circle (Kinsella, 1965). In fact, judging by the most widely used textbooks at the time (and until the late 1950s), the mathematical structure of the geometry course remained stable: different postulational systems, inclusion of newer mathematics, and corrections of flaws in Euclid's *Elements* did not significantly affect the curriculum (Hunte, 1965).

In 1944, the NCTM created the Commission on Post-War Plans, whose mandate was to develop effective secondary programs. This commission sowed the seeds for the reform movement that followed a decade later. Their most distinctive focus was on the idea of differentiation based on need. The commission stated that new and better courses should be provided for the large portion of the school population whose mathematical needs were not met in the traditional sequence of courses. It recommended geometry as the primary 10th-grade course, but only for capable students. For the less capable, a general mathematics course was proposed. It is worth noting that enrollment in geometry had been in steady decline, particularly during the depression years. In 1948–49, only 8.7% of students enrolled in high school (approximately 6 million students) took a plane geometry course, and only 1.4% took solid geometry (Quast, 1968).

The NCTM Commission (1944) also recommended a slight modification in the goals of geometry:

> we should not continue to give so much time to continuous logical development in geometry. Once a student has learned what it means to prove a statement deductively, it is not necessary to devote an entire year to deduc-

> tive proof. Of course, we should continue proofs through the year, and in subsequent years, just as we should continue arithmetic. Meanings and skills are not established once and for all. (p. 209)

In fact, the report went on to state: "there are needs that should be met ... if for no other reason than to help the student become a more effective worker in such school subjects as science and industrial arts" (p. 209). This last statement revealed the commission's bias toward teaching mathematics from a more utilitarian point of view, in order to help students apply the ideas they were learning to the study of other subjects. Kinsella (1965) was sure to include this utilitarian perspective when he summed up the aims of geometry teaching in 1950 as follows: "the opportunity for cultural experience, a love of the subject for its own sake, and the practical use of mathematics in college courses and in some occupations or professions" (p. 12).

During the 1950s, the movement away from "skill and drill" methods became particularly strong. Many educators, particularly in professional journals but also in textbooks, were advocating the "new" idea of leading students to discover facts on their own. For example, Schnell and Crawford's *Plane Geometry: A Clear Thinking Approach* (1953) stated: "geometric relationships take on more meaning when they are discovered by the students than when they are bluntly presented as theorems stated to be proved" (p. ix). According to Kinsella (1965), by 1956 the new emphasis on meaning and understanding had become so strong that some schools were neglecting the development of basic skills. As we shall see, a turning point in the development of the geometry curriculum, in the late 1970s, involved broadening the meaning of the term "basics" so as to include much more than computational skills.

Kinsella's concern regarding a lack of emphasis on basic skills certainly did not apply to Welchons and Krickenberger's *New Plane Geometry* (1956), the most prevalent textbook at the time. However, even in this traditional textbook, one can detect efforts to improve the possibility of student meaning and understanding. In the foreword, the authors listed 17 "Special Features of the Text." That the first feature was "Creation of Interest" is quite telling in that it appeals directly to psychological aspects of geometry education.[13] The textbook, according to the authors, could help secure and hold the pupil's interest because it included illustrations showing applications of geometry to art, nature, industry, and engineering; explanations of the usefulness of geometry as an aid to everyday reasoning; and brief references to the history of geometry. With respect to proofs, the authors mentioned four special features: (1) their gradual approach to formal proof; (2) a special section on the selection of method to be used to prove theorems; (3) a summary at the end of each chapter, on the methods of proof given in the chapter; and (4) the inclusion of the reasons required for constructions given in a two-column format. The final chapter of the book included some additional topics, another special feature, which consisted of applications of geometry to aeronautics, artillery fire, and maps and map reading. In addition, several applications appeared within the content flow of the text. Finally, each chapter included a word list, which students were expected to be able to "spell and use correctly in sentences" (p. 27). This last special feature was to become widespread in geometry textbooks. So while there were few, if any, instances of students discovering facts on their own, the book initiated several provisions meant to support student meaning and understanding.

During the period from 1940 to 1955, debates continued over many of the issues encountered previously: to fuse or not to fuse solid and plane geometry; to continue or not to continue the segregation of algebra and geometry (the United States was, at this point, one of the countries in which segregation persisted—others included Greece, China and Russia); and to teach geometry for its practical benefits, for its improvement of logical thinking, for its scientific approach to problem solving, or for the insights it could provide into mathematics.

While these debates persisted in policy documents and professional journals, a 1949 questionnaire sent to teachers randomly selected from the mailing list of *The Mathematics Teacher* suggested that teachers were more homogeneous—and more traditional—with respect to their own aims. The survey results showed that teachers ranked "To develop the habit of clear thinking and precise expression" highest, with 47% ranking this objective number one (Brown, 1950).

The diverging aims expressed by teachers, on the one hand, and by mathematics educators, on the other, would persist. Usiskin (1969) noted, however, that there was widespread agreement on two goals: "acquisition of geometric facts and understanding of the nature of a deductive system" (p. 12).

1955: BEGINNING OF THE "ERA OF REFORM"

Changes brewing at this time were without precedent in the history of mathematics education, though it must be noted that geometry was the mathematics course least affected. Kinsella (1965) provided a summary of forces behind curriculum reform at the time: the rapid growth of mathematics during the preceding 150 years; the development of science and technology, which necessitated more mathematically trained people; a growing concern over the neglect of higher-achieving students; a historical tendency for college and university materials to move downward into K–12; an increase in collaboration between college and high school levels; the progress of the USSR; the high financial support from federal government in mathematics education; and the emergence of vigorous and imaginative leadership in mathematics education in various university and professional organizations. Indeed, by the early 1960s, many organizations—national, state, and local—were pressing for an updated curriculum.

Many thought that the previous decades, perhaps through the effects of the war, had encouraged mathematics teaching that was excessively utilitarian. These critics thought that mathematics should be taught for mathematics' sake, rather than for any practical applications. Others complained that existing curricula and textbooks contained too little new mathematics, and that the mathematics they did contain was not good mathematics. For example, a typical kind of criticism reflected the Bourbaki fixation with rigor: too many definitions, assumptions, and propositions had been so simplified that, in the words of Wooten (1965), "almost every concept involved is surrounded with a fuzzy aura of uncertainty" (p. 30). Other critics were more concerned with pedagogical issues; for example, Ferguson (1962) felt that geometry must be taught "as a dynamic, growing subject" and that students must be allowed to "create some mathematics, make conjectures and prove them true or false" (p. 16).

In 1954, the Educational Testing Service (ETS) began to study issues in mathematics education, and made its results known to the College Entrance Examination Board (CEEB). In response, the CEEB appointed a Commission on Mathematics to review the existing secondary school mathematics curriculum, and to make recommendations for its modernization. In 1959, 2 years after Sputnik, the CEEB's commission issued one of the most (unexpectedly) influential documents in the history of mathematics education (Fey & Graeber, 2003). The recommendations of the commission included the need for major reforms in the school curriculum in order to bridge the gap between school and college mathematics. These changes were needed so that secondary school curricula would better reflect facets of pure and applied mathematics. Within a very few years, schools across the country reflected the influence of the recommendations through experimental programs and new textbooks. The commission was among the first to propose, at the national level, the substantial changes in high school mathematics that led to, and is now known as, "modern math."[14] The aims for teaching mathematics they supported were as follows:

> Instruction in mathematics designed to meet the needs of secondary school students for general education should aim to teach the students the basic mathematical ideas and concepts that every citizen needs to know, and to explain the essential character of mathematics—how it is used to explore and describe physical reality, and how it is used to contribute through its aesthetic values to one's personal intellectual satisfaction. (1959, p. 11)

Those who went to grade school and high school in the 1960s were the recipients of this New Math; there was a great investment of government money in retraining teachers in using the new curricula, which introduced sets and set operations, mathematical structures, and so on, at every turn. The curricula were largely written by university mathematics faculty. There were strong reactions to the New Math, both from parents and teachers, and also from some in the research mathematics community. One outspoken critic was Morris Kline, who wrote an influential book, *Why Johnny Can't Add: The Failure of the New Math* (1961).

The commission emphasized the importance of understanding the nature and role of deductive reasoning—in algebra as well as in geometry. They also emphasized ideas that were closely connected to the ideals espoused by the Bourbaki Group and the modern mathematics movement, such as an appreciation of mathematics structure ("patterns"), for example; properties of natural, rational, real, and complex numbers; a judicious use of unifying ideas—sets, variables, functions, and relations. With respect to geometry, they promoted the following three aims: (1) the acquisition of information about geometric figures in the plane and in space; (2) the development of an understanding of the deductive methods as a way of thinking, and a reasonable skill in applying this method to mathematics situations, and (3) the provision of opportunities for original and creative thinking by students. These aims were notably silent on the issue of transfer to nonmathematical situations. In fact, the commission wrote, "it is a disservice to the students and to mathematics for geometry to be presented as though its study would enable a student to solve a substantial number of his life problems by syllogistic and deductive reasoning." The Commission pointed out that deductive methods were used, and could be learned, in other courses as well.

Among their recommendations were the following: (1) because of the grave faults in the logical structure of Euclid, "teachers should feel free to modify the Euclidean development" (p. 24) to attain a more incisive and interesting program; (2) there should be a strong intuitive program in grades 7 and 8, which would consist of "the physical geometry of the space in which we live, rather than as an abstract mathematical system"; (3) curricula should include solid geometry as well as other geometries (spherical, projective, and topology); (4) there should be full use of algebraic technique, and a full synthesis of synthetic and analytic means; (5) finally, solid geometry should occupy approximately one-third of the 10th-grade course. The fifth recommendation finally brought about the possibility of fusing plane and solid geometry, which had been recommended by so many previous commissions. In fact, the actual result was the removal of solid geometry as a special course, and the cursory addition of a limited amount of solid geometry—usually in the last chapter of textbooks.

The extent to which teachers adopted the recommendations of the commission could have been predicted by the research conducted by Leisso and Fisher in 1960. They surveyed 186 teachers with 15 questions relating to the commission's suggestions. Most questions received responses that were supportive of the commission's recommendations, but two did not. First, 69% of the teachers felt that the supposed logical deficiencies in Euclid's plane geometry had not been a barrier to student understanding of the logical development of geometry (presumably, they ascribed students' problems to other reasons, which were, unfortunately, not given). Therefore, though they were free to make modifications to the Euclidean development, similar to the ones the Birkhoff and Beatley textbook offered, few teachers actually did. Second, 82% felt that other geometries (non-Euclidean, projective, etc.) should not be taught in school. On the other hand, there was overwhelming support (92%) for the removal of solid geometry as a separate course. Yet, once again, this did not translate into overwhelming support for the fusion of plane and solid geometry.

Around the time of the commission's report, arguments against the use of superposition were revived (recall the earlier NCMR report that recommended against its use), perhaps encouraged by the general sense of discontent with the *Elements*. In particular, two articles appeared to challenge Euclid in *The Mathematics Teacher*: "What is Wrong with Euclid?" (Meder, 1958) and "Why and How We Should Correct the Mistakes of Euclid?" (Daus, 1960). This proved an opportune time for action, particularly for those seeking to avoid superposition in a more mathematically robust way. Long before, Hilbert had developed a postulational system in geometry that took the relation between certain basic figures as undefined, thus removing the need for words such as "coincide," and vague notions such as "same shape" and "same size." He simply assumed one of the triangle congruence theorems, which allowed him to dispose of superposition altogether. So, the solution was waiting for those anti-superposition geometry teachers.

The New Shape of Geometry Textbooks

This period of time marked the onset of many new and influential geometry programs, as the next two sections will document. Four new geometry programs were developed as a result of the commission's recommendations, including the University of Illinois Committee on School Mathematics (UICSM) publication in 1960 and the SMSG geometry program (1961b), funded by the National Science Foundation, which was based on the recommendations of the Commission on Mathematics of the CEEB. Usiskin (1969) noted that the impact of both UICSM and SMSG was powerful: "for the first time large numbers of students in the United States are learning geometry through a system not primarily Euclid's or Legendre's" (p. 28). In fact, Birkhoff's approach, described earlier, gained in popularity, in combination with Hilbert's development, where basic set theory is given.

The SMSG 7–12 program became the most prominent of any other modern program in geometry, even though it had not been intended to be used directly—it was to have provided a model for the development of other textbooks. Its postulational development—a combination of Hilbert's and Birkhoff's—was even adopted by commercial textbooks such as Moise and Downs's *Geometry* (1964).[15] According to Donoghue (2003) SMSG was a success, and this was in part due to the work of Ed Begle, who assembled teams of distinguished mathematicians as well as outstanding high school teachers, and gave them free reign to reimagine the curriculum. One member, Edwin Moise, wrote that the "traditional content of Euclidean geometry amply deserves the prominent place which it now holds in high-school study; and we have made changes only when the need for them appeared to be compelling" (SMSG, 1961a, n.p.). He claimed a "healthy fusion of old and new," explaining that the design of both the text and the problems was based "on our belief that intuition and logic should move forward hand in hand." Indeed, the SMSG materials reorganized the traditional geometry course experience in deductive reasoning and provided a more realistic view of the role played by axioms and formal definitions: assumptions and terms were introduced only when it could be demonstrated that further inference depended on them (Begle, 1974). SMSG also encouraged the trend to "connect up geometry with algebra at every reasonable opportunity, so that knowledge in one of these fields will make its natural contribution to the understanding of both."

Notably, the SMSG geometry materials contained no drawings or photographs, and geometric constructions (with which Euclid's *Elements* began!) did not appear until Chapter 14. Usiskin (personal communication, August 2004) conjectured that constructions were deemphasized for three reasons: (1) the diminishing of the figure and the ancient Greek notion that one can reason from a figure (see Netz, 1998); (2) the highlighting of formal arguments, and the insinuation that constructions are informal; and (3) the increased emphasis on testing and the intrinsic difficulties in having students use rulers and compasses in examination situations. In addition, many constructions in the *Elements* were presented without any mention of purpose or motivation, and the SMSG geometry materials may have decided that such constructions were pedagogically inappropriate. Regardless of the reason, the effect was to move constructions to middle school geometry, which was seen as a more appropriate time for "hands-on" work. The move also had the effect of further divorcing geometry from its foundational tools. Applications were included, but they did not focus on physics alone, which was often used as a mathematical context. William Wooten, member of SMSG (1965, p. 142), aptly explained:

> It was the feeling of most of the writers of the SMSG textbooks that, contrary to the belief of some, not every teenager is fascinated by the behaviors of freely falling bodies or enthralled with phenomena exhibited by gases in closed or leaking containers.

In line with the recommendations of the CEEB, the second edition included analytic geometry, and alternative editions incorporated some vectors and vector techniques.

In addition to its 7–12 program, SMSG developed a K–6 experimental mathematics curriculum, introducing topics such as set theory, informal geometry, probability, algebra, and analysis of number and numeration structures. While these may seem ambitious topics for elementary school students and teachers, the SMSG was responding to Bruner's (1960) suggestion that "any subject can be taught effectively in some intellectually honest form to any child at any stage of development" (p. 33). While some data exist on the use of geometry textbooks at the high school level, both middle and elementary school data is harder to find. There were certainly changes in many of the widely used commercial texts that followed the lead of the SMSG experimental materials, which provided a richer preparation for high school student, as can be seen by state and local school objectives as well as curriculum guides developed throughout the 1960s. But whether the innovations became regular features of classroom instruction and testing remains an open question.

An analysis of National Assessment and commercial standardized test battery syllabi indicated that, in terms of geometry, increased emphasis was being placed on measurement aspects, which were much easier to assess (National Advisory Committee on Mathematics Education, NACOME, 1975). And while "mathematics" was slowly replacing "arithmetic" as most curriculum writers assiduously incorporated topics such as geometry and probability, teachers often skipped these topics in favor of spending more time on computational skills. In a survey of 1,220 teachers conducted by the NCTM (reported in NACOME, 1975), 78% of teachers reported spending fewer than 15 class periods per year on geometry topics. Given that the National Assessment items for 9-year-olds included 25 items on geometry, with half involving recall of names for figures, it seems probable that these 15 classes were devoted to instruction in identification and vocabulary.

The best-selling high school text in the 1960s was Jurgensen, Maier, and Dolciani's *Modern Basic Geometry*, with over 50% of the market share. Both Dolciani and Jurgenson had been members of the SMSG writing team. Moise and Downs's book was the runner-up, and was considered to be more rigorous (for example, it made a distinction between an angle, denoted by "∠ABC," and an angle measurement, denoted by "m∠ABC"). Jurgenson, Donnelly, and Dolciani's *Modern Geometry: Structure and Method* (1963) also followed the general approach of SMSG, but it contained several appealing new features.[16] It was printed as a hardcover book that used multiple color tones and that contained supplementary inserts with transparent overlay pages that produced three-dimensional effects through the use of perspective and red, green, and blue shading. In these, Bhaskara's "Behold!" proof of the Pythagorean theorem was given (see Figure 3.8).

Typical for the era, constructions did not appear until Chapter 10 and solid geometry in Chapter 14. The development of deductive thinking

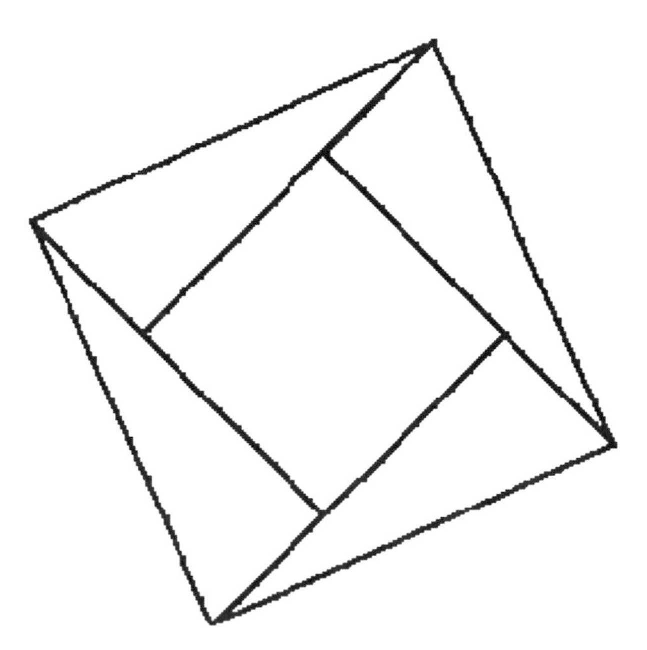

Figure 3.8. Bhaskara's "Behold!" proof of the Pythagorean theorem.

that began in Chapter 3 relied on deductive reasoning encountered in algebra, and theorem demonstrations were formatted so as to require six parts: statement, figure, given, to prove, brief analysis, and statements and reasons in two-column form. For indirect proofs only, a paragraph-form was used. The authors cited two reasons for this departure from the norm: paragraph-form "often seems more natural than a two-column proof" (p. 165) and students need to become familiar with this form because it would be used regularly in mathematics.

Shortly after, the *Unified Modern Mathematics* (Fehr, Fey, & Hill, 1972) curriculum was developed. It attempted to restructure secondary school mathematics from grade 7 to 12, and, according to Donogue (2003), was one of the most thoroughgoing attempts to use an integrated approach—in which all content strands were taught in the same course rather than having them divided into, say, Geometry, Algebra I, Algebra II, and so on. One goal of the new curriculum was to introduce more of the advanced mathematics that the Europeans included in their syllabi. The authors believed this goal could be accomplished by restructuring the curriculum: *Unified Modern Mathematics* "presents the subjects as an integrated body of knowledge, that is one which eliminates the barriers separating the traditional branches of mathematics study, and unifies the subject through study of its fundamental concepts ... and structures" (Fehr, 1970, p. 281). In addition to the reordering, some traditional topics were eliminated, and "clever pedagogical approaches to teaching" were used. Finally, the textbook was taken to be "*the* primary learning material" (p. 291, *italics in original*). Geometry first appeared in Chapter 6, which treated the lattice coordinate system; and Chapter 11 introduced transformations in the plane as well as symmetry.

THE 1960s: TIME FOR TRANSFORMATIONS

In 1962–63, 14% of all high school students and 49% of all 10th-graders were enrolled in a one-year geometry course (Office of Education, 1966). Comparable data in the mid-1950s were 10% and 42%, respectively. It was also estimated, at this time, that 2 million students in grades 7–12 were using SMSG materials (Usiskin, 1969). Comparable statistics are difficult to find in other countries given the fact that geometry study is often integrated with the study of algebra and analysis, which means there are seldom textbooks dealing exclusively with geometry (this has been and continues to be the case in both Canada and the United Kingdom, for example). Abeles (1964) studied the geometry programs of Denmark, England, France, and West Germany, and concluded, "geometry is not taught as a separate subject" (p. 4567). Furthermore, changes in the curriculum content seemed to be accelerating quickly, as both transformations and vector concepts were being increasingly stressed in these countries, while less concern was being placed on "the formal, axiomatic nature of geometry; the synthetic and analytic aspects of geometry are not really integrated until late in the programs" (pp. 4567–4568).

While significant changes were written into the organization of the four new geometry programs that were developed, both the objectives and the content (with the exception of space geometry and analytic geometry) remained untouched. One exception could be found in the work done in 1961, by a group led by Howard Levi (Sitomer, 1964) at Wesleyan University, who began writing a course that treated Euclidean geometry as a special kind of affine geometry (a geometry in which properties are preserved by parallel projection from one plane to another, thus rendering meaningless Euclid's third and fourth postulates). Their goal was to "present geometry as an integral part of contemporary mathematics" (p. 6). They did so by using Levi's coordinate approach, which relied heavily on the algebraic structure of the line to introduce geometry concepts. In fact, one of the staples of the high school geometry course, the perpendicular line, was not introduced in the Wesleyan materials until halfway through the book, in Chapter 7. The book was intended for above-average students. Its approach does not seem to have survived in the curricula of today.

Another group, the University of Illinois Committee on School Mathematics (UICSM), working at about the same time, developed a 2-year course based on the concept of vector space, adding yet another change in the traditional content being offered in geometry textbooks. The authors were influenced by the integrative nature of projective geometry and sought to develop a curriculum based on projective principles:

> beginning with intuitions about parallelism and distance, many intuitive properties of points and translations are observed.... Intuitions about projections lead to the formal concept of the dot product of translations, from which we obtain the geometric notions of perpendicularity and distance. (Szabo, 1966, p. 4)

And still other approaches were developed. One gave considerable attention to topological concepts (see Kelly & Ladd's *Geometry* [1965]) and another began with a distinctive set of undefined terms: point, locus, and grid locus (see Lathrop & Stevens's *Geometry—A Contemporary Approach* [1967]). All these approaches provided evidence of the plethora of ways in which geometry could be conceived, and their proliferation suggested that Euclid's stranglehold was finally waning. While they may have paid little explicit attention to the pedagogical ramifications of using particular approaches, they embodied an intrinsic interest in the logical structures behind different geometries.

The SMSG decided to follow Hilbert's lead on dealing with superposition, in assuming one of the triangle congruence theorems, and also chose to assume properties of the real numbers (SMSG, 1961b). Usiskin (1969) argued that the SMSG definitions of congruence (and similarity) were mathematically precise, but that they lacked "the generality that intuition normally ascribes to the related ideas of 'same shape' and 'same size'" (p. 36). Instead, he, along with Coxford, proposed distance-preserving transformations as an alternative approach, which gave birth to *Geometry: A Transformation Approach* (GATA) (1971). In this approach two figures are said to be congruent if and only if there exists a distance-preserving transformation (such as a reflection, rotation, or translation) that maps one figure onto the other. As Schuster (1967) argued, this approach has logical and aesthetic advantages:

> First of all, it has logical cleanliness. Secondly, it is general, so that the same notion holds for zero-dimension, one-dimension, two-dimension, three-dimensional, and *n*-dimensional configurations. And, thirdly, it is general in the sense that the congruence of general figures is dealt with in a single idea.... The mathematical economy is obvious, and I believe the method has aesthetic appeal as well. (pp. 32–33)

As is often the case with arguments based on aesthetic grounds, disagreements arose over which approach was best. Moise and Downs, in their 1975 revision of *Geometry*, which had been published 7 years earlier, argued against the transformation approach for two reasons: first, they found it more sophisticated than the needs of high school students warranted; and second, they cited the difficulty of finding suitable problems in which transformations could be used—after all, they argued, one of the best features of traditional geometry is the "wealth of interesting problems that the student can solve" (p. vi).

Coxford and Usiskin's (1971) GATA included the first materials developed for the average American high school studying Euclidean geometry that used a "transformation approach." In this approach, definitions for congruence and similarity, given in terms of transformations, are applied to prove the traditional theorems in elementary geometry (see Table 3 for an example). The motivation for developing these materials were threefold: (1) they were deemed more intuitive, thus easing any learning problems that students might have; (2) they possessed mathematical elegance; and (3) they would be relevant to the later mathematics encountered by the student (Usiskin, 1969). In addition to providing a transformation approach to traditional geometry, Coxford and Usiskin emphasized the usefulness of transformations in problem solving. Thus transformations

Table 3.3. A Transformation Proof

Theorem: If two sides of a triangle are equal in length, the angles opposite these sides are congruent.

Let m be the bisector of ∠BCA

A', the reflection image of A, falls on ray CB (angle symmetry)

The reflection image C' of C is C itself (C is on line m)

AC equals A'C' (which equals A'C) (reflections preserve distance)

But AC = BC (given)

So BC = A'C, and since there is exactly one point on a ray at a given distance from the endpoint, A' = B.

Thus, the reflection of ∠ABC over line m is ∠BAC.[a]

The two angles are congruent (definition of congruence)

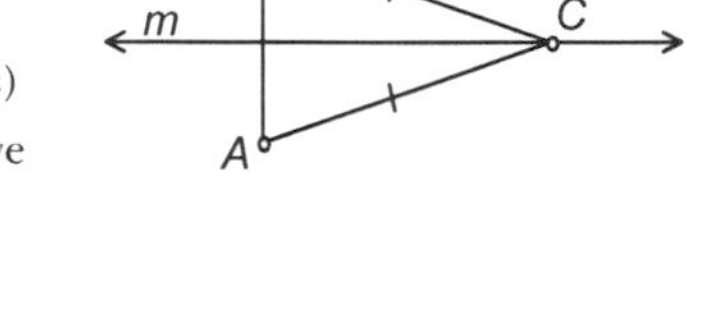

[a]For this, technically, we need to know that reflections preserve betweenness.

became a basis for geometry as well as an important instrument of geometry—on par with the way the ancient Greeks saw locus constructions.

As with SMSG, the GATA materials begin by considering properties of points, lines, planes, angles, angle measure, distance, and betweenness, but then the reflection image of a point is defined and assumed to exist. Following this, two figures are defined to be congruent if and only if one is the image of the other under a reflection or a composition of reflections. Properties of reflections are also brought to bear in proofs. It is important to note that the transformation definition used for congruence and the development of the triangle congruence propositions is followed with an analogous definition for similarity and development of the triangle similarity propositions. The analogical development was an important reason for using the tranformational approach. For other topics, such as circles, areas, and the Pythagorean theorem, traditional approaches were used. In addition to the emphasis on transformations, the importance of visual processes is impressed on the student, with an early chapter on visual experiences. As is discussed shortly, visualization and its relation to geometry would soon become much more prominent in researchers' thinking about the geometry curriculum.

The debate over the many different approaches (vector, transformational, affine) reached its peak in the mathematics education discourse in 1972, when *The Mathematics Teacher* published a series of articles defending the different approaches then being proposed by researchers and textbook writers. Fehr (1972) argued vigorously that the "The present year-long course in Euclidean geometry must go" (p. 151), echoing the words of the mathematician Dieudonné. He proposed three goals of geometry instruction: geometric problem solving, spatial sense, and geometric thinking that is not exclusively formal or deductive. And his conception of geometry was different from, but anticipated by, mathematical developments in the 1980s, particularly in its insistence that geometric and spatial thinking is indispensable to mathematics more broadly. For example, he thought geometry should lead to "a basic understanding of vector spaces and linear algebra" (p. 152).

Such foundational discussions appear rarely in the current educational literature: they seem to be at once too practical to be included in research journals and too theoretical to be included in the current classroom-oriented *The Mathematics Teacher*, which is now more geared to the everyday needs and concerns of high school teachers. A cursory glance at twentieth-century textbooks suggests that the Euclidean approach managed to survive quite strongly, while ideas originally intended to constitute approaches of their own (such as vectors and transformations) have become separate chapters in geometry textbooks, with transformations figuring most importantly in the middle grades.

In addition to changes in curricular approaches brought about in the 1960s, schools had begun altering their course offerings. By 1972, only 48% of United States secondary schools (with enrollment greater than 1,000) offered courses in plane geometry (down from 72% in 1960), while only a paltry 4% offered courses in solid geometry (down from 47% in 1960). On the other hand, 34% now offered a course in plane/solid geometry, up from 0% in 1960 (Ostendorf, 1975).

New Tools for Geometric Manipulation

In addition to changes in curriculum approaches, it would be difficult to ignore the appearance in the 1960s of new artifacts used for teaching geometry, especially at the elementary school level, artifacts that students could use for experimentation and visualization, much as one finds in the science lab. In addition to models such as three-dimensional objects made in wood, plastic, cardboard, wire, or plaster, commercial products designed specifically for geometry learning were invented and developed, such as Mira mirrors used for exploring reflection and translation as well as congruency, tiles that could be used for tessellations, and geoboards. Geoboards were originally made of pins or nails driven into a block of wood in a rectangular array (see Figure 3.9), and were developed by the well-known English mathematician and educator Caleb Gattegno in the early 1950s. In addition to the models and manipulatives devoted to geometry, other manipulatives such as Dienes blocks and Cuisinaire rods

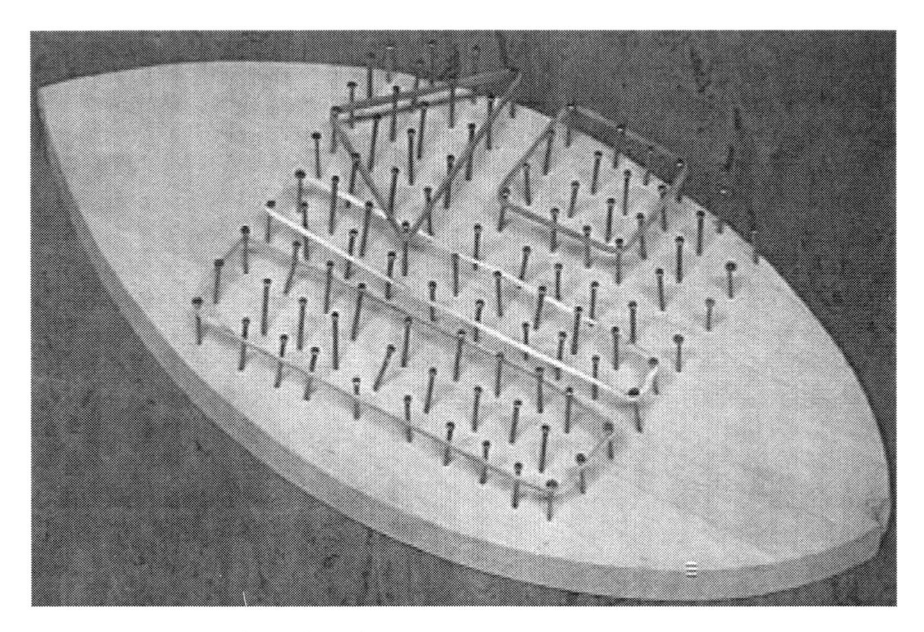

Figure 3.9. An early geoboard.

began to appear, providing teachers with geometric ways of introducing algebraic ideas and concepts.[17]

The interest in manipulatives, with their focus on hands-on, visual learning, focused largely on elementary school mathematics. However, as we shall see, the appeal of visualization started to spread up through the grades in the late 1960s with the advent of animated films and videos, which provided a natural segue to the computer technology that would begin to flourish in the 1980s.

1975: THE NACOME REPORT REDEFINES "BASIC SKILLS"

During the early 1970s, mathematical competence was still viewed, and perhaps even relatively more so given the supposedly more educated public, as being equivalent to computational ability. Several factors contributed to this, including (1) a decline on standardized achievement tests and college entrance examinations; (2) reactions to the poor results of the National Assessment of Educational Progress (NAEP); (3) the rising costs of education and increasing demands for accountability; (4) the shifting emphasis from curriculum content to instructional methods and alternatives; (5) the increased awareness of the need to provide remedial and compensatory programs; and (6) the widespread publicity given to each of the above by the media. The National Institute of Education (NIE) responded to these pressures by adopting the area of basic skills as a major priority and sponsoring the Conference on Basic Mathematical Skills and Learning, held in Euclid, Ohio, in October 1975.

Seventy-five educators were invited to the conference, and were asked to make suggestions as to the course of action that should be taken to address the pernicious issue of basic skills, whose narrow interpretation by the public threatened the health of the mathematics curriculum. Due to the political and ideological firestorm caused by the ambitious social studies program developed by Jerome Bruner in 1972/73, *Man: A Course of Study* (MACOS), the NSF had been essentially forced out of curriculum-related initiatives, and the NIE was interested in figuring out where to place their future efforts. The growth of educational technologies at the time was signaled by Seymour Papert's suggestion, which was that the NSF should purchase calculators for every student (handheld calculators had come out in 1971). However, the most influential result of the meeting was the formation of the National Advisory Committee on Mathematical Education (NACOME), which reported on the state of mathematics education in their 1975 publication *Overview and Analysis of School Mathematics Grades K-12*, widely known as the NACOME report.

The report's discussion of computer and calculator technology was quite striking. It noted that 58% of American secondary schools made use of computers at the time, with 43% of instructional uses in mathematics. While the primary effect of these uses of computers was to boost student motivation and interest, the report called attention to the changes in curriculum that should take place as a result of the presence of the new technologies. In fact, the report ascribed the need to reconsider "basic skills" due to the increasingly widespread availability of technology that can perform many of the functions previously taken to be basic. Computers were also seen as "beginning to demonstrate new ways of approaching traditional equation solving problems and the symbolic manipulation that require so much instructional time in secondary school." Thus, instead of emphasizing computational ability, the report shifts focus to skills that would be needed to work with the new technologies, skills such as approximation, orders of magnitude, and interpretation of numerical data. The applications discussed by the NACOME report were focused on areas of mathematics such as computation, algebra, and trigonometry, and ignored the Logo programming language (even though Papert was one of Logo's developers). Logo had already been tested with second-, third-, and seventh-graders in 1968–69, and had focused on the development of formal thinking—a topic much closer to the geometry curriculum. Indeed, it was not until the invention of turtle graphics and the evolution of the personal computer in the 1980s that Logo, which is now well known for its geometric applications, began the eventual impact of computing technology on the geometry curriculum.

The NACOME report paved the way for new recommendations from school mathematics, beginning with the NCTM *Agenda for Action* in 1980. It propagated, as well, the trend of separating geometry from measurement, two content areas that had often been treated together in textbooks and policy recommendations. The NACOME report argued vigorously for doing away with the debilitating dichotomies flourishing in the "new math" versus "old math" debates that had begun in the "era of reform," and asserted that the notion of basic skills—which was at the time, once again, making a comeback, as a backlash to the perceived failure of the "New Math" movement—should be expanded far beyond previous conceptions, which stressed arithmetic and computation. Despite being held in Euclid, Ohio, the NIE conference list of 10 basic skills did not include geometry, though it did include items such as problem solving, estimation and approximation, and computer literacy. This seems to have been an oversight as the follow-up position paper on basic skills in mathematics, published by the National Council of Supervisors of Mathematics (NCSM) in 1978, which borrowed heavily from the NIE conference report—using identical verbiage in many parts—did include geometry as well as measurement as two of the basic mathematical skills. The NCSM list of 10 basic skills needed by students who "hope to participate successfully in adult society." These basic skills formed the basis for the list that would eventually become adopted by many groups, including the Third International Mathematics and Science Study (TIMSS), NAEP, and the NCTM *Standards*.

With geometry as a basic skill that all students should have, the ground was set to begin its downward stretch into the elementary school grades. In addition to the influence of the NCSM position paper, Usiskin (personal communication, August 2004) saw the ensuing entry of geometry into the K–5 years as being supported both by the explosion of workshops offered at NCTM conferences, in which hands-on activities involving geometry were ideally suited, and by the concrete manipulative movement, which supported geometric investigations particularly well.

The NCSM published its report in *The Mathematics Teacher* in 1978. Following in the footsteps of the NIE report, it stated that "[l]earning to solve problems is the principal reason for studying mathematics" (p. 148). Mathematical skills were seen as falling under 10 vital areas: problem solving; applying mathematics to everyday situations; alertness to the reasonableness of results; estimation and approximation; appropriate computations skills; geometry; measurement; reading, interpreting, and constructing tables, charts, and graphs; using mathematics to predict;

and computer literacy. Interestingly, while the NIE report omitted geometry, the NCSM position paper omitted algebra.

With respect to geometry, the NCSM position paper stated that:

> Students should learn the geometric concepts they will need to function effectively in the 3-dimensional world. They should have knowledge of concepts such as point, line, plane, parallel, and perpendicular. They should know basic properties of simple geometry figures, particularly those properties that relate to measurements and problem-solving skills. They also must be able to recognize similarities and differences among objects. (1978, p. 149)

The position paper made no mention of proof or logical reasoning; the focus was on the kinds of geometric concepts that could begin to develop in the elementary grades. As we will see, several mathematicians and educators later began lamenting the loss of the geometric visualization and intuition that had been banned by the Bourbaki Group, and argued for a wider conception of geometry than the one based solely on proof.

It is interesting to consider the period from 1975 to 1980—one of incipient shifts in the United States—in other countries, where important changes had also taken place. For example, Laborde (1995) tracks the shifts over this short period of time by comparing French textbooks published in 1975 with others published in 1979. Four changes were very evident: (1) the increased use of specific or numerical examples; (2) the increased presence or acceptance of diagrams and drawings; (3) the inclusion of exercises involving the use of instruments such as rulers and protractors; and (4) the appearance of new kinds of problems. In the 1975 books, no measures were given, and, if it happened that numbers were needed, they were denoted by letters. No mention was ever made of the possibility of drawing a diagram, and no construction exercises involving instruments were included.

In contrast, many of the exercises in the 1979 textbooks specified the length of segments, using numbers, and either implicitly or explicitly called for a drawing. The later editions also included construction exercises using instruments. Finally, several new kinds of problems were included, such as the determination of the trajectory of a point depending on a moving point or a moving line, the computation of inaccessible lengths (such as the height of a tree using similar triangles), and optimization problems. In general, these problems were much more contextualized than had been their predecessors. While similar changes would occur in the textbooks of the United States, they would be much more gradual, owing, in part, to the absence of directing national standards.

THE 1980s: THE DE-DEGRADATION OF "GEOMETRIC CONSCIOUSNESS"

In 1974, mathematician Philip Davis made the following plea to his fellow mathematicians: "visual geometry ought to be restored to an honored position in mathematics" (p. 113). Davis based his argument partially on the possibilities for geometric visualization offered by the color and animation capacities of computer technology. Five years later, he and a colleague, James Anderson, augmented their plea, again relating it to the computer: "Restore geometry. Restore intuitive and experimental mathematics. Give a proper place to computing and programmatics. Make full use of computer graphics" (1979, p. 125). They lamented the phenomenon (documented earlier in this monograph) that had made visualization fall into disrepute, dubbing it the "degradation of the geometric consciousness."

Over the course of the ensuing decade, several other voices and movements, some but not all of them originating in the world of computers, began to make a strong case for the importance of visualization in mathematics and for the role of geometry in supporting and nurturing visual and intuitive modes of reasoning. In the preface to his book *Geometry and the Imagination*, Hilbert and Cohn-Vossen (1952/1983) had already set the stage for this development:

> In mathematics ... we find two tendencies present. On the one hand, the tendency toward abstraction seeks to crystallize the *logical* relations inherent in the maze of material that is being studied, and to correlate the material in a systematic and orderly manner. On the other hand, the tendency toward *intuitive understanding* fosters a more immediate grasp of the objects one studies, a live rapport with them, so to speak, which stresses the concrete meaning of their relations.... With the aid of visual imagination [*Anschauung*] we can illuminate the manifold facts and problems of geometry, and beyond this, it is possible in many cases to depict the geometric outline of the methods of investigation and proof.... In this manner, geometry being as many faceted as it is and being related to the most diverse branches of mathematics, we may even obtain a summarizing survey of mathematics as a whole, and a valid idea of the variety of its problems and the wealth of ideas it contains. (p. iii)

Many other mathematicians and educators joined Hilbert and Davis and Anderson in their quest to restore geometric visualization. By 1980, mathematics educator Caleb Gattegno was able to observe that "[geometry] is coming back because the mathematicians who are not rabid algebraists have realized that they need some substance" (p. 15). He noted the impossibility of truly escaping geometry, by observing that even the most abstract algebraic notions are described in geometric metaphors (the more present-day work of Lakoff and Nuñez [2000] supports this claim as well). He might also have made the argument that geometry actually represents other mathematics; for example, graphs and diagrams, which are geometrical ideas, are used in most areas of mathematics, including algebra, analysis, and statistics. It may well be the computer that, with its increasingly accessible ways of displaying and manipulating visual information, will provoke the comeback of geometry. In fact, visual displays have already become ubiquitous, and have had an impact on the teaching of algebra, as I discuss later. Pursuing this idea of the pictorial nature of mathematics, Gattegno proposed that mental imagery should be taken as the "stuff" of geometry, partly in an attempt to distinguish or even reclaim geometry from the increased encroachment of algebra in the school mathematics curriculum.

The renewed interest in visualization resulted in a 1991 publication by the Mathematical Association of America titled *Visualization in Teaching and Learning Mathematics* (edited by Zimmerman and Cunningham). The editors described visualization as "the process of producing or using geometrical or graphical representations of mathematical concepts, principles or problems, whether hand drawn or computer generated" (p. 1). They emphasized their case for the importance of visualization by citing evidence showing that students are generally lacking in visual geometry skills (see, e.g., Brown et al., 1988) and argued that such skills are essential to many branches of mathematics, as well as to "crystallographers, biochemists, surgeons, aviators, mechanical shovel operators, sculptors, choreographers and architects"[18] (Baracs, 1980, p. 4). More recently, in 2001, the renowned mathematician Sir Michael Atiyah (2001) also linked geometry to visualization and intuition when he wrote:

> spatial intuition or spatial perception is an enormously powerful tool and that is why geometry is actually such a powerful part of mathematics.... We try to put [things] into geometrical form because that enables us to use our intuition. Our intuition is our most powerful tool. (p. 50)

The computing powers cited by Davis and Anderson in 1979 were, at the time, not yet available to students; however, by the early 1980s, the development of the microcomputer and the emergence of the turtle graphics version of the programming language Logo finally put some of this promising new technology in the hands of students and teachers. In Logo turtle geometry, the student controls the movement of the turtle (an on-screen cursor) by inputs of various commands involving moving forward, backward, and turning. Papert (1980) called the geometry of the turtle "body-syntonic" in the sense that geometric shapes were created in reference to the student's own body, instead of in reference to a Cartesian coordinate system, or an extrinsic point of view. Solving a problem involved walking through it as if one were the turtle. The programming component of Logo enabled students to define new commands using Logo primitives and previously defined commands, and then combine them in a logical sequence, thus effectively "teaching" Logo how to draw any geometric shape. The benefits of Logo were only partially related to the narrow sense of geometry learning, and extended to, for example, problem-solving strategies and metacognitive processes such as reflective processing.

Researchers conducted several studies using Logo, with a wide range of results. Clements and Battista (1989), for example, showed that students' conceptualization of angles, shapes, and motions were uniformly higher for a Logo group as compared to a control group. On the other hand, students' problem-solving abilities were not necessarily improved by the use of Logo (Battista & Clements, 1986). Nevertheless, Logo was incorporated into several elementary and middle school curricula, particularly in support of the learning of geometry. For example, the first edition of the NSF-funded *Connected Mathematics Project* (Lappan, Phillips, Fey, & Fitzgerald, 1996) included a geometry investigation to explore shapes and transformation using a commercially available Logo-like software program. And while the presence of Logo has considerably waned in North America, new versions of it, which offer several more powerful features, continue to be widely used in other countries such as the United Kingdom and Brazil.

Several factors contributed to the decline of Logo in geometry curricula in the United States. A cluster of factors relates to the perpetual difficulties teachers and schools had (and continue to have) in integrating computer-based technology. Cost, time, and lack of expertise were frequently cited constraints. With regard to the geometry curriculum itself, the use of any new tool, whether it be computer-based or not, changes the kind of geometry that can be done, and the ways in which students can learn geometry. As a result, adopting a program such as Logo involved changing the questions, activities, and assessment strategies teachers were accustomed to using—an endeavor that required both mathematical and pedagogical expertise. An additional barrier presented by Logo relates to its distance from the geometry curriculum, if only in terms of vocabulary and actions. Words such as "square," "circle," "rotate," and "radius" do not appear anywhere in the primitives of Logo. Instead, teachers and students found commands such as forward, left turn, and repeat. This made the match between curricula and textbooks on the one hand, and the technology on the other, more difficult for many teachers to make (see Sinclair & Jackiw, 2005).

Despite the eventual waning of Logo in the United States, its short-lived success and excitement may have helped jump-start the interest in geometry, particularly at the lower grades, and may have also paved the way for the interest in and acceptance of next-generation software programs for geometry learning, such as the *Geometry Supposer* (Schwartz & Yerushalmy, 1985). This construction program allowed students to choose a primitive shape, such as a triangle or parallelogram, and perform measurement operations and geometric constructions on that shape. The sequence of constructions was recorded by the program and was automatically performed on other shapes, thereby allowing students to make and test conjectures such as, Does the sum of the angles of a triangle remain constant? Do the diagonals of the parallelogram bisect each other?

Research showed that students were better able to identify nonstereotypical examples of shapes (such as a triangle or square whose base is not horizontal) and to extend their learning beyond traditional geometry content, for example, formulating conjectures and posing problems (Clements & Battista, 1992). The emergence of this software for geometry learning also required researchers to consider how to motivate for students the need for proof. When measurement-based evidence is available, as well as inductively generated inferences based on checking a property for many cases, students were left wondering why proofs were needed at all. Of course, motivating the need for proof had long been a concern in geometry education, but the introduction of software for geometry learning such as the *Supposer*—and later dynamic software programs that are described shortly—seem to have helped mathematics educators identify and describe the very different roles that proofs can play in mathematical inquiry, and also acknowledge the importance that conviction plays in one's decision to develop a proof.

Trends in Courses Offered and in the Geometry Education Literature

The burgeoning interest in visualization, in geometric thinking, and in computer-based explorations of geometry, combined with the inclusion of geometry as a basic skill in the NCSM report, proved to be quite persuasive. The impact can be seen in NCTM'S third volume on geometry education, titled *Learning and Teaching Geometry, K–12*, published in 1987. By analyzing the titles of all three NCTM volumes on geometry (shown in Table 3.4), one can trace the changes in emphases that transpired over the preceding 50 years: The title of the 1930 publication suggests that geometry was segregated from the rest of mathematics; the 1973 publica-

Table 3.4. Comparison of Sections in the Three NCTM Yearbooks on Geometry

The Teaching of Geometry (1930)	*Geometry in the Mathematics Curriculum (1973)*	*Learning and Teaching Geometry, K–12 (1987)*
Informal versus intuitive geometry	Informal Geometry K–12	Perspectives on Geometry
Demonstrative versus deductive geometry	Formal Geometry in the Senior High School	A View of Problem Solving and Applications
Plane versus solid geometry	Contemporary Views of Geometry	Activities in Focus
Formal versus informal proof	The Education of Teachers	A Look at Geometry and Other Mathematics
		Preparing Teachers

tion established a connection between geometry and the rest of the mathematics curriculum; and the 1987 publication suggests that geometry learning begins in elementary school. The 1987 publication also reflected the increasing number of advocates to the view that geometrical concepts can be used to model problems and phenomena in other domains.

The Teaching of Geometry was briefly mentioned earlier, and the four entries in the table describe the prevalent issues discussed in its pages. *Geometry in the Mathematics Curriculum* contained 15 essays that signal an important change in trends in the teaching of geometry from the 1930 publication. Part I focused on the need to include a substantial amount of informal geometry at the K–6 level, perhaps as a result of the modern mathematics movement. The question was no longer "whether informal geometry" but "how early should we begin teaching geometry." Part II did away with the debates of the 1930 volume—solid versus plane, Euclidean versus non-Euclidean, or informal versus demonstrative—and instead focused on the new choices that have emerged as viable during the intervening years, including synthetic geometry, coordinate geometry, transformational geometry, affine geometry, integrated geometry, and even "eclectic" geometry. Part III analyzed modern geometry and its implications for geometry education. And, finally, Part IV addressed the education of teachers, a topic that was absent in the 1930 publication.

Learning and Teaching Geometry, K–12 contained 20 essays and included several new emphases that revealed an interest in the questions of how geometry is best taught and how a student actually learns geometry. Indeed, the first article in Part I provided an overview of the van Hiele model of the development of geometric thought, which would later become the basis for several textbooks (including Hoffer's [1979] textbook mentioned earlier, which has been the high school textbook most faithful to the van Hiele theory), achievement frameworks, as well as the subject of much educational research (see, e.g., Burger & Shaughnessy, 1986; Mayberry, 1983; Senk, 1989). The van Hiele model (1986) is based on five levels of geometric understanding:

- Level 0: Visualization. Students are aware of space, but only as something that exists around them.
- Level 1: Analysis. Through observation and experimentation students begin to discern the characteristics of figures.
- Level 2: Informal Deduction. Students can establish the interrelationships of properties both within figures and among figures.
- Level 3: Deduction. The significance of deduction as a way of establishing geometric theory with an axiomatic system is understood.
- Level 4: Rigor. Non-Euclidean geometries can be studied and different systems can be compared.

Many researchers have questioned these levels, both in terms of their supposed linearity and in terms of their helpfulness in improving student understanding (see, e.g., Gutierrez, Jaime, & Fortuny, 1991). Nevertheless, they have provided a strong and sustained basis on which to predict, characterize, and compare students' geometric understanding.

In Part II, the focus turned to problem solving and applications, and to efforts to create real-world problems that can be solved with the use of geometric concepts. Part III contained five articles, all illustrating various "hands-on" approaches to geometry teaching, with titles such "Visualizing Three Dimensions by Constructing Polyhedra" and "Conic Sections: An Exciting Enrichment Topic." In Part IV, the four essays made connections to other parts of mathematics (such as probability, number theory, and calculus), enhancing the 1973 publication's emphasis on geometry being a part of the whole mathematics curriculum. Finally, Part V built a case for the importance of teacher training in geometry.

Reversing Trends in Students Taking Geometry

In an earlier section, I documented the decline of geometry—a decline that began with a decreased emphasis at the research mathematics level, and then seeped into undergraduate programs and then high schools. The renewal of geometry that the preceding paragraphs indicate seems to have eventually reached the high school level, reversing prior trends. Looking at the statistics over the period from 1980 to 2000, we find a

Table 3.5. Percent of Public High School Graduates Taking Algebra I and Geometry in Grades 9–12

Course	*1982*	*1987*	*1990*	*1994*	*1998*	*2000*
Algebra I	55.2	58.8	63.7	65.8	62.8	61.7
Geometry	47.1	58.6	63.2	70.0	75.1	78.3

marked and steady increase in the percentage of public high school graduates taking geometry courses (National Center for Education Statistics, 2003). The percentage of high school graduates taking Algebra I increased less consistently and significantly in the same time period. This may be explained by the NAEP data, which show that 33% of students reported taking algebra in eighth grade—and this course does not show up in high school transcripts.

Recent data can be found in the International Congress on Mathematics Education (ICME)-10 report conducted by Dossey and Usiskin, called *Mathematics Education in the United States 2004—A Capsule Summary*. In that report the authors documented that 94% of high school graduates have taken a year of algebra and 88% have taken geometry. Today, geometry is still not as prominent at universities, both in terms of the research interests of faculty and in terms of courses offered at the undergraduate and graduate levels. It seems that other supporting forces were at play in reviving geometry during the last two decades of the twentieth century, the computer being an important one—as we shall see.

1989 NCTM: *CURRICULUM AND EVALUATION STANDARDS FOR SCHOOL MATHEMATICS*

As the NACOME report admitted, "fundamental decisions on curriculum content and instructional practice are traditionally made at the local school level—usually allowing classroom teachers substantial freedom in choice of mathematics and methods" (p. xi). The United States does not have an official national mathematics curriculum. However, in 1989, the NCTM published its *Curriculum and Evaluation Standards for School Mathematics*. The *Standards* document had been more than 10 years in the making, and its appearance had an important effect on the mathematics education community both in the United States and elsewhere. The *Standards* positioned itself as the mathematics education community's response to the call for reform in the teaching and learning of mathematics; "they reflect, and are an extension of, the community's responses to those demands for change" (p. 1). The mandate of the *Standards* was to improve *all* of mathematics education with the inclusion of *all* students.

The reasons given by the *Standards* for teaching geometry were inherent to the nature of geometry, contrasting significantly with previous policy statements on the goals and nature of geometry education. Usiskin's reasons for teaching high school geometry, made in 1980, anticipated many of the views present in the 1989 *Standards*: "1. Geometry *uniquely* connects mathematics with the real physical world. 2. Geometry *uniquely* enables ideas from other areas of mathematics to be pictured. 3. Geometry *nonuniquely* provides an example of a mathematical system" (p. 418). Indeed, the *Standards* emphasized, at the grades 9–12 level, the wide applicability of geometry to recreations, practical tasks, the sciences, and the arts. In the aims for the grades 5–8 level, the connection to the physical world and to other areas of mathematics was also made: "geometry helps students represent and make sense of the world. Geometric models provide a perspective from which students can analyze and solve problems, and geometric interpretations can help make an abstract (symbolic) representation more easily understood" (p. 112). And finally, the stated aims for geometry at the K–4 levels began "geometric knowledge, relationships, and insights are useful in everyday situations and are connected to other mathematical topics and school subjects. Geometry helps us represent and describe in an orderly manner the world in which we live" (p. 48). The *Standards*' view of geometry can even be seen in the other content strands, where visual displays of mathematics ideas are widely emphasized and sometimes even dominant: graphs of functions, probability trees, addition and multiplication grids, area models for multiplying binomials, and many more.

The specific standards relating to geometry education for grades 9–12 emphasized the following aspects of geometry: integration across topics at all grade levels, coordinate and transformation approaches, the development of short sequences of theorems, deductive arguments expressed orally and in sentence or paragraph form, computer-based explorations of two-dimensional and three-dimensional figures, three-dimensional geometry, and real-world application and modeling. Many previously emphasized ideas in earlier documents and earlier textbooks received decreased attention in the 1989 *Standards*: Euclidean geometry as a complete axiomatic system, proofs of incidence and betweenness theorems, geometry from a synthetic viewpoint, two-column proofs, inscribed and circumscribed polygons, and theorems for circles involving segment ratios. Part of the vision of the *Standards* was that justification and reasoning become matters for all students, in all areas of mathematics—not only in geometry. This took some of the pressure off geometry, and allowed educators to focus on other worthy and integral aspects of geometry, such as visualization.

Early Examples of Standards-Based Curricula

The University of Chicago School Mathematics Project (UCSMP), which began work in 1983, produced a series of six textbooks for grades 7–12. These textbooks, which became popular in the 1990s, were described as "the first full mathematics curriculum to implement the recommendations of the *NCTM Standards* Committee" (1990, p. T5). For the 1998–99 school year, UCSMP (1998–99) estimated that "over three million students [used] UCSMP elementary and secondary materials" (p. 6). In its introductory notes to the student, the geometry book (authored by Coxford, Usiskin, and Hirschhorn) characterized the subject as "the study of visual patterns" (p. v). Algebra concepts and skills were used throughout to "motivate, justify, extend, and help students with important geometry concepts" (p. iv). The textbook was designed to be consistent with van Hiele's (1986) theory of geometric understanding. It also included lessons that *required* the use of calculators or computers; teachers were even urged to allow students to use calculators on geometry tests.

The UCSMP materials introduced at least one pedagogical innovation: SPUR was the acronym for the four dimensions of understanding that students were supposed to achieve: Skills (drawing and visualizing, in addition to following algorithms), Properties (understanding properties, mathematical relationships, and proofs), Uses (applications of geometry), and Representations (graphic approaches such as coordinate systems, networks, and diagrams). Geometric tools were introduced early on (in Chapter 3), followed by transformations in Chapter 4, making and testing conjectures in Chapter 5, and formal proof was delayed until Chapter 6. The textbook emphasized visualization and the improvement of drawing skills, and even included step-by-step instructions for sketching solids in perspective.

Michael Serra's *Discovering Geometry* was another high school textbook that was widely seen as implementing the recommendations of the *NCTM Standards.* First published in 1989, the book drew on the development of geometric thinking in adolescents (including the van Hiele theory), and provided many opportunities for students to "create geometry for themselves" (p. iii). It also included Logo computer activities. Formal proofs were delayed until the last two chapters, at which time students were to have had visual, analytic, and inductive experiences with many geometric concepts. Investigations and cooperative learning were particularly stressed. Still today, revised versions of this textbook accounts for approximately 10% of the high school market.

Prior to the *Standards*, Porter (1989) reported that the major focus of most elementary and middle school geometry was on recognizing and naming geometric shapes, writing the proper symbolism for simple geometric concepts, developing skill with measurement and construction tools such as the compass and protractor, and using formulas in geometric measurement. If some fourth- and fifth-grade teachers reportedly spent "virtually no time teaching geometry" (p. 11), many others were left to teach a "hodgepodge of unrelated concepts with no systematic progression to higher levels of thought" (Clements & Battista, 1992, p. 422). The 1989 *Standards* affirmed the importance of studying geometry in grades K–8 and proposed approaches that addressed the findings of recent research, as well as the developmental levels of students in these grades. Students would begin with hands-on experiences, with vocabulary growing out of experience and understanding, instead of being taught first and foremost. This approach was intended to help alleviate unnecessary memorizing, and to prepare students for the importance of and need for vocabulary. By the later middle school grades, students would be able to organize information about the shapes they had observed and begin to see, for example, that a square has properties of both the rhombus and the rectangle.

At the time of the writing of the *Standards*, the UCSMP had already begun research and development on an elementary school curriculum that adopted the same approach as the *Standards*; it was later published as *Everyday Mathematics*. In the early 1990s, several other elementary curricula (including *Investigations in Number, Data, and Space*), middle school curricula (including *Mathematics in Context* and *Connected Mathematics Program*), and high school curricula (including *Contemporary Mathematics in Context, Interactive Mathematics Program*, and *Math Connections*) went into development, with funding for design and field-testing by the NSF; they were also aimed at implementing the 1989 *Standards* document. Many of these initial curricula have now revised their first edition textbooks and have conducted research showing that students using these curricula perform as well if not better than students using non-reform curricula (see Senk & Thompson, 2003). The high school–level *Interactive Mathematics Program* (IMP) is particularly interesting in terms of some of the debates that have been raised over the course of this monograph, since it integrates different content strands within single units. For example, the "Shadows" unit, in which students develop formulas for finding the length of a shadow, integrates geometry and algebra.

1991: DYNAMIC GEOMETRY SOFTWARE

In the 1940s, Jean-Louis Nicolet and Caleb Gattegno began to create films[19] that can be seen as early precursors of what has come to be known as "dynamic geometry." In one film ("Circles in the Plane"), a red circle appears on the screen and moves about, appearing to be shrinking and growing, or perhaps to be moving nearer and further away. In a later sequence of the film, a circle passes through two fixed points. The circle grows bigger until it becomes a straight line and then shrinks on the other side of the two points. The early films used very simple combinations of points, lines, and circles to evoke relationships that remain invariant when these objects moved while being subject to various constraints. Gattegno (1989) commented on the importance of the visual experience these films gave students:

> we have indicated how we educate the whole being by letting images and their dynamics be the forerunners of words. Verbalization about one's experience after seeing a film is valuable. But, it can only translate in a limited way the ineffable experience of infinite classes of simultaneous impressions gathered from the animated entities on the screen. Infinity is now actual; it is only latent when language is used. (p. 153)

A half-century later, new mathematics software would not only provide compelling visual experiences for students, it would put students in control of their experience, able to grow and shrink circles at their own pace, in their own time.

The beginning of the 1990s saw the forceful and influential emergence of a new kind of software—today, collectively called dynamic geometry software[20] (DG)—that took some of the features of the Geometry Supposer (Schwartz & Yerushalmy, 1985), such as construction and measurement, but that offered entirely new, dynamic representations. The software provided primitive mathematical objects, such as points, lines and circles, along with basic tools for acting on these objects, such as "construct midpoint," "translate line" and "measure radius." Figure 3.10 shows a configuration between three basic objects, circle, point, and segment. As the point on the circle travels around the circle, the point on the segment, which travels back and forth along the segment, traces out the red "flower." In marked contrast to Logo, for example, these objects and tools spoke the language of the school geometry curriculum.

While DG environments have given rise to many interesting issues in geometry education, and now, across the curriculum, their introduction and widespread popularity (primarily at the high school level) has had several discernible effects on many of the issues touched on in this monograph. In addition to raising questions about the aims of teaching and

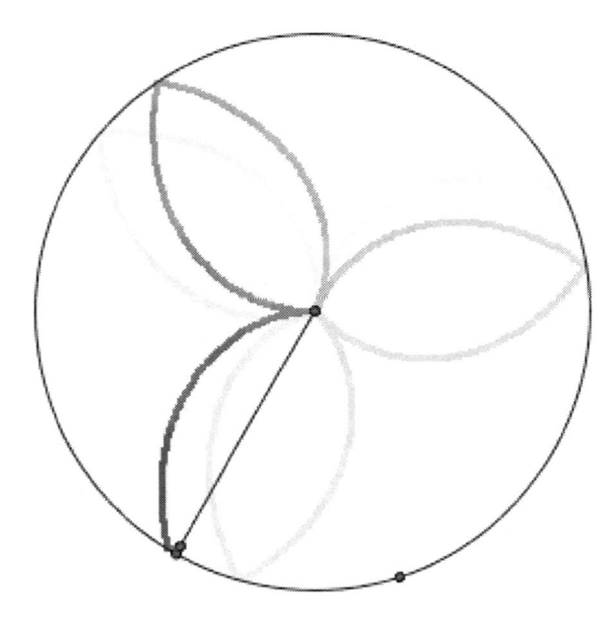

Figure 3.10. A dynamic geometry construction in The Geometer's Sketchpad.

learning geometry, DG environments have irrevocably authorized the use of *motion* in geometry—the dragging of objects is its characteristic function. The introduction of motion into previously static diagrams or processes has reacquainted geometry with *time*, so that constructions and transformations have a linear ordering that can now be sped up, slowed down, or even viewed in reverse, depending on the desire of the user. And by introducing a new set of more powerful instruments (not only a virtual compass and straightedge, but also tools such as "perpendicular line," which, though constructible by compass and straightedge, are now available as primitives), DG environments have reinstated the fundamental role of tools in geometry (prompting, more recently, much research on the hitherto unexplored processes by which people come to adopt and deploy these tools—see, e.g., Artigue, 2002). In "Charting a New Course for Secondary Geometry," Chazan and Yerulshamy (1998) emphasized the way in which dynamic geometry software has changed the curricular representation of how mathematics is done, both by connecting construction and deduction with construction problems, and by introducing conjecturing and proofs of student conjectures.

With regard to the aims of geometry education, the presence of DG software has supported an increased emphasis in the curriculum on the processes of experimentation and exploration, as well as an earnest attempt to develop better opportunities for students to engage in mutually supporting inductive and deductive modes of reasoning. One can also discern a renewed interest in developing in students an intuitive sense of the behaviors of mathematical objects, and in delaying the point at which the algebraic formulation of phenomena occurs. For example, Goldenberg, Cuoco, and Mark (1992) wrote:

> In turning [a geometrical theorem] into a dynamic experiment, students not only develop a deeper insight into the result and its explanation; they also discover ways to tinker with the experiment in ways that suggest new results and new explanations. At the same time they are developing a sense for continuous functions that makes intuitive use of the delicate properties of the real numbers in ways ... that have become lost in present day curricula that concentrate on algebraic formalisms. (p. 28)

DG environments seem to exploit Tahta's (1980) way of describing geometry, as "an awareness of imagery that arises from a dynamic process of the mind" (p. 4) as opposed to algebra, which he construes as "the formalization of such awareness."

The affordances of the software, coupled with the constructivist leanings of many using the software, gave rise to a plethora of new "discovery learning" classroom activities in which students constructed certain configurations, such as a triangle, and were led to make conjectures about relationships in these configurations—for example, that the sum of the angles of the triangle is always 180°, regardless of the type of triangle one has. While some of these activities were possible in nontechnology settings, the speed of computation and the range of examples (hundreds of triangles can be effortlessly tested) offered by DG environments has made them more compelling. However, their very compellingness led many teachers and researchers to question the role of proof in the curriculum, since conviction can be obtained quickly and relatively easily by dragging. Early on, Yerushalmy, Chazan, and Gordon (1993) found that DG environments could often prevent students from understanding the need and function of proof.

Prior to the arrival of DG environments, teachers were urged to impress upon students the limitations of empirical methods: it looks like the three corners of the triangle can be turned and made from a straight line, but how do we know the line is perfectly straight? By appreciating these limitations, students would feel more compelled to seek deductive methods. Alternatively, students were shown pictures of optical illusions and reminded that what they saw could not always be trusted: more secure methods of arriving at knowledge were necessary. While similar methods could be used in DG environments (Sketchpad reports the sum of the angle measures as 180.0000° but perhaps the next decimal place will be a 4!), they were much less successful in promoting the need to prove. After all, students see computers as being reliable and exact. Furthermore, the use of empirical methods should be coordinated with the use of deductive methods, not set in opposition to them.

This situation has led researchers such as de Villiers (1990, 1997, 1999), who has worked extensively with *The Geometer's Sketchpad*, to articulate somewhat novel ways to promote proof. Instead of questioning the conviction empirical methods can give, de Villiers invited students to accept the evidence, but then to ask *why* the relationships they are seeing must hold. Of course this approach will not be novel to mathematicians, who have quite often discussed the role of proof in their own work as providing insight and explanation rather than conviction. For example, mathematician George Pólya (1954) famously claimed that mathematicians never begin working on a proof unless they are convinced of its truth: "Without such confidence we would have scarcely found the courage to undertake the proof.... When you have satisfied yourself the theorem is true, you start proving it" (pp. 83–84).

The presence and promise of DG environments has contributed to tension in the discourses on the aims of geometry, as attested to by the renewed interest in and debates about geometry. For example, in the early 1990s, the International Commission on Mathematics Instruction (ICMI) became "worried about the future of Geometry teaching" (Griffiths, 1998, p. 194) and commissioned a study that was published in 1998, entitled *Perspectives on the Teaching of Geometry for the 21st Century*. Griffiths characterizes the tension as a struggle between the "top-down pressure" of "Theorem-proving Mathematicians" (or "TPMs," p. 194) with the "bottom-up pressure" of mass education systems. The TPMs advocate a certain kind of geometric fluency, which Griffith characterized as (1) understanding theorems of continuous mathematics in terms of pictures rather than symbol manipulation; (2) converting the sketches into the formal language of a "proof" in the style of Euclid; and (3) acquiring a taste for the discipline of organizing a theory, with Euclid as a paradigm (p. 195). In contrast, the "bottom-up" pressure involved addressing more practical needs such as measuring dwellings for quantities of materials, understanding plans and maps, and having a spatial sense. Griffiths concluded that curriculum developers are thus under increased pressure to choose interesting topics (see also Goldenberg, 1989).

While DG software has been used extensively in the high school grades, its use in the middle school grades has only quite recently increased (see Battista, 2002), while its use in the primary grades is just beginning to emerge (see Sinclair & Crespo, 2006). While changes at the elementary school have always been slower, the case in geometry, and thus in dynamic geometry, had to wait for the relatively recent widespread and sustained inclusion of geometry in the primary curriculum. In 1998, Lehrer, Jenkins, and Osana conducted a research study on elementary school students' understanding of space, taking their cue from the NCTM's (1991) recommendations that geometry instruction should begin in the primary grades. As predicted by van Hiele (1986), most children's geometric reasoning about two-dimensional form (in tasks such as identifying shapes) most often involved the visual appearance of figures. However, their reasoning seemed to be very amenable to motion: Lehrer and colleagues reported that some children "regarded figures as malleable objects that could be pushed or pulled to transform them into other figures, without any concern for the effects of these transformations on the properties of the figures" (p. 142). So a chevron and a triangle were alike because "if you pull the bottom [of the chevron] down, you can make it into this [the triangle]" (p. 142). More than 20% of the children provided morphing explanations of similarity. The researchers suggested that the limited number of and canonical type of examples in textbooks may contribute to children's slow development in reasoning about two-dimensional form. It will be interesting to observe the impact that the use of dynamic geometry software will have on young learners' geometric thinking. This area of research is bound to grow in the upcoming years as more resources are developed for the elementary school curriculum.

2000: NEW AND IMPROVED *PRINCIPLES AND STANDARDS*

In 2000, the NCTM published a thorough revision of its 1989 *Standards* publication, called the *Principles and Standards for School Mathematics 2000*. The document was intended to be a resource and guide for all who made decisions that affect the mathematics education of students in pre-K through grade 12. The recommendations in it were grounded in the belief that all students should learn important mathematical concepts and processes with understanding. The content skills included the now-familiar quintet: Number and Operations, Algebra, Geometry, Measurement, Data Analysis and Probability. The process skills include Problem Solving, Reasoning and Proof, Communication, Connections, and Representation.

Regarding geometry, the *Principles and Standards* stated that all students should (1) Analyze characteristics and properties of two- and three-dimensional geometric shapes and develop mathematical arguments about geometric relationships; (2) specify locations and describe spatial relationships using coordinate geometry and other representational systems; (3) apply transformations and use symmetry to analyze mathematical situations; and (4) use visualization, spatial reasoning, and geometric modeling to solve problems. Again, contrary to the traditional use of proof only in high school, the *Principles and Standards* proposed that students of all ages engage in justifications of the claims of others, whether a conjecture about a property, or simply providing the rationale for an answer. In a statement summing up this position, the document declared "Reasoning and proof should be a consistent part of students' mathematical experience in pre-kindergarten through grade 12" (p. 56).

It is interesting to compare the 1989 and 2000 documents in terms of specific objectives at the different grades levels. The more narrowly focused grade-level categorization of the 2000 document (K–2, 3–5, 6–8, and 9–12) makes direct comparison to the 1989 document (K–4, 5–8, and 9–12) somewhat difficult. The 2000 document included much more detail about specific expectations than did the 1989 one, which also makes comparison more difficult. However, some changes can be detected. In terms of content at the grades K–2 levels, the 2000 document included three-dimensional geometry as well as the recognition and application of flips, turns, and slides. It also mentioned the study of perspective and symmetry. At the grades 3–5 level, the specific expectations emphasized the development of vocabulary to describe the attributes of two- and three-dimensional shapes, and placed increased importance on the making and testing of conjectures about geometric properties and relationships as well as the development of logical arguments to justify conclusions. Students were also expected to recognize geometric ideas and relationships and *apply* them to other disciplines and to problems that arose in the classroom or in everyday life. Finally, the representation of three-dimensional objects received increased attention.

At the grades 6–8 level, the use of coordinate geometry was emphasized, and more rigorous terms appeared in the description of the expectations. See the comparison in Table 3.6 for examples (especially italicized words). The 2000 document again included a reference to deductive thinking in this grade band, in the context of the study of the

Table 3.6. Comparing the 1989 and 2000 *Standards* for Middle School Geometry

1989: Grades 5–8	*2000: Grades 6–8*
Explore transformations of geometric figures.	Describe *sizes, positions, and orientations* of shapes under informal transformations such as flips, turns, slides, and scaling.
Represent and solve problems using geometric models.	Use geometric models to represent and *explain* numerical and algebraic relationships.
Identify, describe, compare, and classify geometric figures.	*Precisely* describe, classify, and understand relationships among types of *two- and three-dimensional objects using their defining properties.*

Table 3.7. Comparing the 1989 and 2000 *Standards* for High School Geometry

1989: Grades 9–12	*2000: Grades 9–12*
Classify figures in terms of congruence and similarity and apply these relationships.	Explore relationships (including congruence and similarity) among classes of two- and three-dimensional geometric objects, *make and test conjectures* about them, and *solve problems* involving them.
Deduce properties of, and relationships between, figures from given assumptions.	*Establish the validity of geometric conjectures using deduction, prove theorems, and critique arguments made by others.*
Interpret and draw three-dimensional objects.	*Analyze properties and determine attributes* of two- and three-dimensional objects.

Pythagorean theorem: "create and critique inductive and deductive arguments concerning geometric ideas and relationships, such as congruence, similarity, and the Pythagorean relationship" (p. 397).

The most salient change at the grades 9–12 level, at least organizationally, involved the combination of the two standards "Geometry from a synthetic perspective" and "Geometry from an algebraic perspective" into one unified standard. Many of the similar expectations appeared, though their wording again reflected more rigor and precision—as Table 3.7 illustrates (see, especially, italicized statements)—as well as increased attention to the making, testing, and critiquing of conjectures. In each of *Standards* 7 and 8, the 1989 document included arguably more rigorous statements directed to "college-intending" students: "develop an understanding of an axiomatic system through investigating and comparing various geometries" and "deduce properties of figures using vectors." In contrast, the 2000 document made no special mention of college-intending students, thereby intending its expectations for *all* students.

The changes articulated between the 1989 and 2000 *Standards* documents represent rather small differences, particularly in comparison with the changes that have occurred since the turn of the twentieth century. After having lost its central place in the curriculum during the middle and later parts of the century, geometry had to struggle to reestablish its place, and this it did by moving far beyond the strict Euclidean tradition to which it was bound at the beginning of the century.

NOTES

1. Legendre's text had an extremely long life, going through numerous editions; an American text by Davies based on it was still in use in American universities in the early twentieth century (Quast, 1968).

2. The full gamut of shortcomings was not known even to Legendre. They were not appreciated until near the end of the nineteenth century, when mathematicians realized the implications of non-Euclidean geometries. That Hilbert, one of the greatest mathematicians of his time, took it upon himself around the turn of the twentieth century to try to work out a postulate system for Euclidean geometry, indicates how long it took for this realization.
3. The point is made humorously by the famous geometer H.S.M. Coxeter (1961) when he describes the game of Vish. This game involves beginning with a specific word (such as "point") and looking it up in the dictionary. The definition will include other words that are unknown (such as "position" and "magnitude"), and must also be looked up. Eventually, one encounters a vicious circle in which one word is used again to define another unknown word. As Coxeter writes, "The only way to avoid a vicious circle is to regard certain primitive concepts as being so simple and obvious that we agree to leave them undefined" (p. 6).
4. This aesthetic goal was begun by Euclid, who proved his first 27 propositions using only Postulates 1–4; he held off using Postulate 5 as long as possible. As a result, those first 28 propositions are true in the non-Euclidean hyperbolic and elliptic geometries.
5. It should be noted that while several changes were proposed to Euclid's *Elements*, these changes all respected the five Euclidean postulates so that the term Euclidean geometry is not just confined to the specific geometry given in Euclid's *Elements*, but also to revised forms that yield all of the theorems that are in Euclid's *Elements*.
6. Note that spherical geometry on the whole sphere does not satisfy the axiom that two lines can intersect in at most one point.
7. There is yet another type of displacement that could be mentioned, which has been ignored, that could be used to describe the phenomenon of dilation (size transformation). Dilation is as fundamental to similarity as rigid motions are to congruence, and certainly involve some kind of motion.
8. Usiskin and Griffin (2007) find another significant impact of the NCMR report, in terms of the definitions of quadrilaterals. The report recommended inclusive definitions of rhombus, square, etc., and after 1930 one does not see exclusive definitions. Also one does not see rhomboid and trapezium. However, the definition of trapezoid remained exclusive.
9. Interestingly, *Humanized Geometry* begins with a section on "how to measure an angle," which shows students how to use a protractor and leads them to the idea that no measurement can be considered to be correct. The lack of dependability of measurement continues to be a popular method for introducing the importance of proof and construction, especially with the current widespread use of dynamic geometry software.
10. Loci were already struggling for a place in the curriculum. As early as 1934, McLeod published an article in *The Mathematics Teacher* titled "The Use of Locus Problems in the Teaching of Geometry," in which he argues for the merit of locus problems. Nonetheless, loci were a fixture in the curriculum in the 1940, 1950s, and even the 1960s.

11. There were several others before, one of the most notable being Felix Klein's Erlanger program. This significant influence is described in the section "1902: Perry and Moore and Theories of Learning."
12. The SMSG books adopted two of Birfkhoff and Beatley's postulates, the "Ruler Postulate" and the "Protractor Postulate," which made it possible to use properties of real numbers in the logical structure of the geometry.
13. Popkewitz (2004) has argued that developments in school curricula have continually and increasingly contributed to the "alchemy" of disciplinary subjects, a process through which psychological concerns take precedence over disciplinary ones. This process of alchemy seems to have been slower in geometry than in other subject areas.
14. Other, more local, committees preceded the CEEB's: the University of Illinois Committee on School Mathematics (the first of the New Math projects) had, by 1959, developed an entire curriculum for grades 7–10 with a new geometry course. SMSG started in 1957 and before the CEEB report had delineated many of the ideas that would become known as the New Math. The 1959 NCTM yearbook (planned at least a year earlier) already reflected the ideas of the CEEB report. The CEEB report, then, could be seen as instantiating what was already being done by projects into a national report.
15. Note that Moise himself had written the SMSG geometry.
16. There are at least three geometry texts with Dolciani's name. The best-seller of the 1960s was *Modern Geometry: Structure and Method* (Jurgenson, Donnelly, & Dolciani, 1963). A very short time later, Houghton Mifflin came out with *Modern Basic Geometry* for slower students that deemphasized proof. Later, they had two proof-oriented geometry (and corresponding algebra) texts.
17. An early proponent of these materials and their use in elementary mathematics education was Lore Rasmussen of the Miquon school, who created the first "mathematical laboratory" in the United States in the early 1960s.
18. That list could easily be extended: in addition to medical imaging and global positioning systems, for example, the computer animation industry makes extensive use of sophisticated geometry, and has been steadily increasing since the 1995 release of the movie *Toy Story.*
19. The College Geometry Project later developed animated films on topics such as kaleidoscopes and isometries in the late 1960s that were intended to improve high school and college geometry teaching.
20. The term *dynamic geometry* was originally coined by Nicholas Jackiw and Steven Rasmussen, but has quickly become a generic term used to characterize the now-many software packages that allow for the continuous, real-time transformation known as "dragging," including *The Geometer's Sketchpad* (Jackiw, 1991), *Cabri-Géomètre* (Baulac, Bellemain, & Laborde, 1988), *Cinderella* (Kortenkamp & Richter-Gebert, 1999), *Geometry Inventor* (Brock, Cappo, Dromi, Rosin, & Shenkerman, 1994), and, in a partial way, *Super-Supposer* (Schwartz & Yerulshalmy, 1993).

CHAPTER 4

WHERE ARE WE NOW AND WHERE ARE WE GOING?

Quast (1968) ended his overview of the history of the geometry curriculum by commenting on the unprecedented changes brought about in schools as a result of the post-1950s reform period. Although changes in content were limited, he noted the "significant changes in attitude" that were evident in the eclectic view of the mathematics curriculum as well as in teachers' tendency to teach "enthusiastically and for understanding" (p. 326). The proliferation of alternative programs and the continued development of experimental programs were signs that uniform reform strategies of the past had finally ceded to more diverse initiatives. The development of the NSF-funded curriculum projects (known as the reform curricula) combined the two trends in offering several alternative programs that met more or less uniform reform strategies, with different programs emphasizing different strategies.

One trend Quast remarked upon was the increased involvement of curriculum developers in implementing changes in the classroom. This has certainly continued in the years since then, with new programs being introduced by curriculum developers working in close association with teachers and schools. Quast wrote that the textbook played an important role in the evolution of geometric teaching, but lamented the continued demand of teachers for nonreform textbooks. The situation seems to have improved cautiously, with increased numbers of students learning in class-

The History of the Geometry Curriculum in the United States
pp. 91–94

rooms that use reform-based materials. However, the pressures of the No Child Left Behind (NCLB) Act and of the Math Wars[1] have taken away much of the initiative and choice that teachers had just two decades ago. Writing today, Quast would have to lament the demands of government officials, parents, special-interest groups, and administrators. It seems that the availability of exemplary curricula has grown in proportion to the political complexities involved in making the decisions that will result in individual teachers or students being able to use these curricula. The times when teachers argued over the number of propositions a geometry textbook should have or over the number and difficulty level of "originals" are long past and may seem, in comparison, much more manageable.

In terms of aims, Quast found that the two aims associated with earlier times—practical reasons and transfer of training to life situations—ceded to more mathematical aims such as understanding geometric facts and the role of deductive thinking. He noted that the methods made available to teachers, whether through textbooks, professional development, or instruments, to meet new aims often lagged behind their appearance in policy documents and even textbooks, as was the case with the emerging emphasis on student discovery as a way to develop insight and understanding of geometric principles. As mentioned before, the current aims of geometry education, as expressed in policy documents such as the *Principles and Standards* 2000, combine González and Herbst's (2004) four discourses (mathematical arguments, formal arguments, intuitive arguments, and utilitarian arguments). This may succeed in meeting the demands and expectations of various stakeholders; however, it makes for very complex, and sometimes contradictory, guiding values. This situation is endemic to education as a whole, and unlikely to change in the near future.

With respect to content, Quast found very little change between 1890 and 1965, with only minor modifications in the postulates, some decrease in the number of basic theorems, and some increase in the number of exercises. Since Quast's writing, it is safe to say that numerous changes have taken place. While textbooks have not, in general, adopted approaches completely distinct from Euclid's, they have embraced vectors, transformations, and algebraic methods in their development. However, few students have the opportunity to learn about non-Euclidean geometries—or fractal geometry—during their K–12 mathematics program, despite the existence of several high-quality supplementary resources. The incipient movement to offer extensive geometry in the primary grades has certainly continued to gain momentum and the study of both solid and plane figures has been realized. On the other hand, less progress has been made in teaching geometry in a more integrated manner, particularly at the high school level.

It is unlikely that Quast could have anticipated the profound effect that computer-based technology has had on the geometry curriculum. Barring similarly unexpected and significant influences, what trends might we reasonably look forward to in the next 20 years? Once again, it is worth looking back in time. In his overview of articles in *L'Enseignement Mathématique* related to geometry at the turn of the twentieth century, Bkouche (2003) highlights the emergence of two important, renewing themes related to the teaching of geometry: fusion and motion. Both themes have been discussed throughout the monograph, and have important roots in ancient Greek practices. In addition, both can be seen as having contributed to the renewal of the teaching of geometry we have witnessed in the United States over the past few decades. Motion seems to have been accepted in the geometry curriculum for many of the same reasons given by mathematicians and has been particularly well suited to developments in visualization and computer-based technologies. However, fusion has succeeded less for its relationship to developments in projective geometry than for its affinity with other aims and developments in geometry education, such as providing young students with informal experiences of geometry relevant to the three-dimensional world in which they live. From these two examples, we might infer that changes in the geometry curriculum must have support from mathematicians and research mathematics, but must not necessarily be adopted for mathematical reasons.

Another theme that has contributed to the renewal of geometry education is related to the methods of geometry, and, in particular, to the emerging empiricist conception of geometry (which, incidentally, reconnects geometry with its dual nature, as a branch of mathematics as well as physics). Several factors have contributed to this conception, from developments in mathematics related to the devaluation of the *Elements*, to shifting priorities in the aims of geometry education. It will be interesting to see whether the empiricist conception continues or grows.

While documents such as the NCTM *Principles and Standards 2000* provide a comprehensive overview of the aims and methods agreed upon by many different stakeholders in mathematics education, they do not necessarily provide information about the typical geometry curricula most students in the United States encounter. Throughout this monograph, I have joined other historians of the geometry curriculum in relying on policy documents, textbooks, research articles, and various statistics to chart changes in the *intended* curriculum. One might expect a different picture of the *implemented* curriculum experienced by students. This would vary widely depending on local factors such as the textbooks and supplementary materials a given teacher chooses (and it is important to note that the reform-based textbooks described above are by no means the most widely used ones) or the amount of professional development a given school offers as well as on broader factors such as the customized standards and large-scale assessment tools of a given state. While it is tempting to want to describe the *actual* geometry curriculum, its sheer diversity would make the task virtually impossible. Furthermore, doing so would neither negate nor replace the many interesting changes we have been able to follow in the discourse around the geometry curriculum since the beginning of the nineteenth century.

NOTE

1. The current "new" Math Wars echo the earlier argument over the "New Math" in the 1960s and 1970s. The old math wars were centered on the introduction of "New Math," but also involved arguments about whether or not the old ways were failing children. The current wars grew in response to the NCTM *Standards* and to arguments about what counts as "basic" in mathematics learning, and about the relative importance of teaching computational skills, algorithms, and conceptual understanding,

REFERENCES

Abeles, F. (1964). College preparatory programs in geometry of four nations: A critique for the study of U.S.A. programs. *Dissertation Abstracts, 25*, 4567–4568.

André, P., & Lormeau, E. (1908). *Géométrie*. Paris: E. André Fils.

Artigue, M. (2002). Learning mathematics in a CAS environment: The genesis of a reflection about instrumentation and the dialectics between technical and conceptual work. *International Journal of Computers for Mathematics Learning,* 7(3), 245–274.

Atiyah, M. (2001). Mathematics in the 20th century: Geometry versus algebra. *Mathematics Today, 37*(2), 46–53.

Austin, C. M. (1919). *A history of plane geometry as a school study in the United States.* Unpublished master thesis, University of Chicago.

Baracs, J. (1980). Geometric perception of space. *Structural Topology, 4*, 4.

Battista, M. (2002). Learning geometry in a dynamic computer environment. *Teaching Children Mathematics, 8*(6), 333–339.

Battista, M., & Clements, D. H. (1986). The effects of logo and CAI problem-solving environments on problem-solving abilities and mathematics achievement. *Computers in Human Behavior, 2*(3), 183–193.

Baulac, Y., Bellemain, F., & Laborde, J. M. (1988) *Cabri-géomètre*, un logiciel d'aide à l'enseignement de la géométrie, logiciel et manuel d'utilisation. Paris: Cedic-Nathan.

Beatley, R. (1935). Third report of the committee on geometry. *Mathematics Teacher, 28*, 329–380.

Begle, E. (1974). Review of "Why Johnny Can't Add." *National Elementary Principal, 53*, 26–31.

Bell, E. T. (1940). *Men of mathematics*. New York: Dover.

Betz, W. (1930). The transfer of training, with particular reference to geometry. *The teaching of geometry. Fifth yearbook of the National Council of Teachers of Mathe-*

matics. New York: Bureau of Publications, Teachers College, Columbia University.
Betz, W., (1933). *Junior mathematics for today*. Boston, New York: Ginn & Company.
Betz, W. (1936). The reorganization of secondary education. *The place of mathematics in modern education. Eleventh yearbook of the National Council of Teachers of Mathematics*. New York: Teachers College, Columbia University.
Billingsley, H. (1570). *The elements of geometric of the most ancient philosopher Euclid of Megara*. London: Citizen of London.
Birkhoff, G., & Beatley, R. (1940). *Basic geometry*. Boston: Scott, Foresman, and Company.
Birkhoff, G. D. (1932). A set of postulates for plane geometry, based on scale and protractor. *Annals of Mathematics*. *33*(2), 329–345.
Bkouche, R. (2003). La géométrie dans les premières années de la revue. In D. Coray, F. Furinghetti, H. Gispert, B. Hodgson, & G. Schubring (Eds). *One hundred years of l'Enseignement Mathématique: Moments of mathematics education in the twentieth century* (pp. 95–112). Geneva: L'Enseignement Mathématique.
Blackhurst, J. H. (1935). *Humanized geometry: An introduction to thinking*. Des Moines, Iowa: University Press.
Bonnycastle, J. (1798). *Elements of geometry, containing the principal propositions in the first six, and the eleventh and twelfth books of Euclid*. London: J. Johnson.
Bourlet, C. (1928). *Cours abrégé de géométrie*. Paris: Hachette.
Breslich, E. (1933). Secondary school mathematics and the changing curriculum. *Mathematics Teacher, 26*(6), 327–349.
Breslich, E. (1951). How movements of improvement have affected present day teaching of mathematics. *School Science and Mathematics, 51*(445), 131–141.
British Mathematical Association (1923). *Teaching of geometry in schools*. Glasgow, UK: Glasgow University.
Brock, C. F., Cappo, M., Dromi,D., Rosin, M., & Shenkerman, E. (1994). *Tangible math: Geometry inventor*. Cambridge, MA: Logal Educational Software and Systems.
Brown, C., Carpenter, T., Kouba, V., Lindquist, M., Silver, E., & Swafford, J. (1988). Secondary school results for the fourth NAEP mathematics assessment: Algebra, geometry, mathematical methods, and attitudes. *Mathematics Teacher, 81*(5), 337–347.
Brown, K. (1950). Why teach geometry? *Mathematics Teacher, 43*(3), 103–106.
Brown, J. (1999). *Philosophy of mathematics: An introduction to the world of proofs and pictures*. New York: Routledge.
Bruner, J. (1960). *The process of education*. New York: Vintage Books.
Burger, W., & Shaughnessy, J. (1986). Characterizing the van Hiele Levels of development in geometry. *Journal for Research in Mathematics Education*, *17*(1), 31–48.
Butterworth, B. (1999). *What counts: How every brain is hardwired for math*. New York: The Free Press.
Byrne, O. (1847). *Euclid's elements*. London: William Pickering.
Campbell, W. (1899). *Observational geometry* (Phillips-Loomis Mathematical Series).
Chasles, M. (1989). *Aperçu historique sur l'origine et le développement des methods en géométrie*. Paris: Gabay. (Original work published 1837)

Chauvenet, W. (1898). *Treatise on elementary geometry.* Philadelphia: Lippincott. (Original work published 1887)

Chazan, D., & Yerushalmy, M. (1998). Charting a course for secondary geometry. In R. Lehrer & D. Chazan (Eds.), *Designing learning environments for developing understanding of geometry and space*. Mahwah, NJ: Erlbaum.

Christofferson, H. C. (1938). Geometry a way of thinking. *Mathematics Teacher, 4*, 147–155.

Clairaut, A. C. (1741). *Élémens de géometrie*. Paris: Chez David.

Clark, J. R., & Otis, A. S. (1925). *Plane geometry* (Experimental ed.). New York: The Lincoln School of Teachers' College.

Clark, J. R., & Otis, A. S. (1927). *Modern plane geometry.* Yonkers-on-the Hudson, NY: World Book.

Clements, D. H., & Battista, M. T. (1989). Learning of geometric concepts in a logo environment. *Journal for Research in Mathematics Education, 20*(5), 450–467.

Clements, D. H., & Battista, M. T. (1992). Geometry and spatial reasoning. In D. A. Grouws (Ed.), *Handbook of research on mathematics teaching and learning: A project of the National Council of Teachers of Mathematics*. New York: Macmillan.

Commission on Mathematics. (1959). *Program for college preparatory mathematics*. New York: College Entrance Examination Board.

Commission on Post-War Plans of the NCTM. (1944). *The Role of mathematics in consumer education*. Washington, DC: The Consumer Education Study.

Committee on Secondary School Curriculum of the Progressive Education Association. (1940). *Mathematics in general education, Report of the Committee on the Function of Mathematics in General Education*. New York: Appleton-Century-Crofts.

Coolidge, J. (1963). *A history of geometrical methods*. New York: Dover.

Coxeter, H. S. M. (1961). *Introduction to geometry.* New York: Wiley.

Coxford, A. F., & Usiskin, Z. (1971). *Geometry: A transformation approach*. River Forest, IL: Laidlaw Bros.

Crozet, C. (1817). *A treatise on descriptive geometry for the use of the cadets of the United States military academy.* New York: A. T. Goodrich and Co.

Daus, P. (1960). Why and how we should correct the mistakes of Euclid? *Mathematics Teacher, 53*(8), 576–81.

Davis, P. (1974). Visual geometry, computer graphics and theorems of perceived type. *Proceeding of Symposia in Applied Mathematics.* Providence, RI: American Mathematical Society.

Davis, P., & Anderson, J. (1979). Nonanalytic aspects of mathematics and their implication for research and education. *SIAM Review, 21*(1), 112–117.

Davison, C., & Richards, C. (1907). *Plane geometry for secondary schools*. Cambridge, UK: University Press.

Decker, F. (1912). Educational values of geometry. *Mathematics Teacher, 5*(9), 31–35.

De Villiers, M. (1990). The role and function of proof in mathematics. *Pythagoras, 24*, 17–24.

De Villiers, M. (1997). The role of proof in investigative, computer-based geometry: Some personal reflections. In J. King & D. Schattschneider (Eds.), *Geometry turned on* (pp. 15–24). Washington, DC: Mathematics Association of America.

De Villiers, M. (1999). *Rethinking proof with Sketchpad*. Emeryville, CA: Key Curriculum Press.

Dienes, Z. P., & Golding, E. W. (1967). *Geometry through transformations*. New York: Herder & Herder.

Donogue, E. (2003). Algebra and geometry textbooks in twentieth century America. In G. Stanic & J. Kilpatrick (Eds.), *A history of school mathematics* (Vol. 1, pp. 329–398). Reston, VA: National Council of Teachers of Mathematics.

Dossey, J., & Usiskin, Z. (2004). *Mathematics education in the United States 2004: A capsule summary fact book written for ICMI-10*. Reston, VA: NCTM.

Egan, K. (2002). *Getting it wrong from the beginning: Our progressivist inheritance from Herbert Spencer, John Dewey, and Jean Piaget*. New Haven, CT: Yale University Press.

Eisenstein, E. (1983). *The printing revolution in early modern Europe*. Cambridge, UK: Cambridge University Press.

Euclid's *Elements*. (c. 300 BCE). Translated with introduction and commentary by Sir Thomas Heath, 2nd edition. New York: Dover, Vol. I, 1956.

Eves, H. (1972). *A survey of geometry* (Revised ed.). Boston: Allyn & Bacon.

Fawcett, H. P. (1938). *The nature of proof. Thirteenth yearbook of the National Council of Teachers of Mathematics.* New York: Bureau of Publications, Teachers College, Columbia University.

Fehr, H. (1953). Theories of learning related to the field of mathematics. *Twenty-first yearbook of the National Council of Teachers of Mathematics*. Washington, DC: NCTM.

Fehr, H. (1970). The secondary school mathematics curriculum improvement study: Goals—the subject matter—accomplishments. *School Science and Mathematics, 70*, 281–291.

Fehr, H. F. (1972). The forum: What should become of the high school geometry course? The present year-long course in Euclidean geometry must go. *Mathematics Teacher, 65*(2), 102, 151–154.

Fehr, H., Fey, J., & Hill, T. (1972). *Unified mathematics, course I*. Menlo Park, CA: Addison-Wesley.

Ferguson, W. (1962). Pedagogical reasons for the innovations. *Report of an orientation conference for geometry with coordinates.* Stanford, CA: School Mathematics Study Group.

Fey, J. T., & Graeber, A. O. (2003). From the New Math to the *Agenda for Action*. In G. Stanic & J. Kilpatrick (Eds.), *A history of school mathematics* (Vol. 1, pp. 521–558). Reston, VA: NCTM.

Freeman, H. (1932). *An elementary treatise on actuarial mathematics* (2nd ed.). Cambridge, UK: Published for the Institute of Actuaries at the University Press.

Gardner, H. (1993). *Multiple intelligences: The theory in practice*. New York: Basic Books.

Gattegno, C. (1980). On the foundations of geometry. *For the Learning of Mathematics, 1*(1), 10–16.

Gattegno, C. (1989). *The science of education part 2B: The awareness of mathematisation*. New York: Educational Solutions.

Gergonne, J. D. (1825–1882). Philosophie mathématique. Considérations philosophiques sur les élémens de la science de l'étendue. *Annals of Math, 16*, 209–231.

Goldenberg, P. (1989). Seeing beauty in mathematics: Using fractal geometry to build a spirit of mathematical inquiry. *Journal of Mathematical Behavior, 8*, 169–204.

Goldenberg, P. Cuoco, A., & Mark, J. (1992). *Making connections with geometry.* Paper presented at the Geometry Working Group, ICME Québec.

González, G., & Herbst, P. (2006). Competing arguments for the geometry course: Why were American high school students to study geometry in the twentieth century? *International Journal for the History of Mathematics Education, 1*(1), 7–33.

Greenleaf, B. (1858). *Elements of geometry with practical applications to mensuration.* Boston: Robert S. Davis & Co.

Griffiths, H. B. (1998). The British experience. In C. Mammana & V. Villani (Eds.), *Perspectives on the teaching of geometry for the 21st century: An ICMI Study.* The Hague, The Netherlands: Kluwer Academic.

Gutierrez, A., Jaime, A., & Fortuny, J. (1991). An alternative paradigm to evaluate the acquisition of the van Hiele levels. *Journal for Research in Mathematics Education, 22*(3), 237–251.

Hadamard, J. (1947). *Leçons de géométrie élémentaire* (2 volumes). Paris, France: Armand Colin. (Original work published 1898)

Hartshorne, R. (2000). Teaching geometry according to Euclid. *Notices of the AMS, 47*(4), 460–465.

Heath, T. L. (1926). *Euclid: The thirteen books of the Elements* (Vol. I). Cambridge, UK: Cambridge University Press.

Hedrick, E. R. (1917). *Constructive geometry.* New York: Macmillan.

Herbst, P. G. (2002). Establishing a custom of proving in American school geometry: Evolution of the two-column proof in the early twentieth century. *Educational Studies in Mathematics, 49*(3), 283–312.

Hilbert, D., & Cohn-Vossen, S. (1983). *Geometry and the imagination* (P. Nemenyi, Trans.). New York: Chelsea.

Hill, G. (1887). *Geometry for beginners.* Boston: Ginn & Co.

Hill, T. (1854). *First lessons in geometry.* Boston: Ginn & Co.

Hlavaty, J. (1950). *Changing philosophy and content in tenth year mathematics.* New York: Teachers College, Columbia University.

Hoffer, A. (1979). *Geometry: A mode of the universe* (Teacher's ed.). Reading, MA: Addison-Wesley.

Hunte, B. (1965). Demonstrative geometry during the twentieth century: An account of the various sequences used in the subject matter of demonstrative geometry from 1900 to the present time. *Dissertation Abstracts, 26*(January), 3979.

Jackiw, N. (1991). *The Geometer's Sketchpad.* Berkeley, CA: Key Curriculum Press.

Jurgensen, R. C., Donnelly, A. J., & Dolciani, M. P. (1963). *Modern geometry: Structure and method* (Teacher's ed.). Boston: Houghton Mifflin.

Jurgensen, R. C., Donnelly, A. J., & Dolciani, M. P. (1972). *Modern school mathematics geometry.* Boston: Houghton Mifflin.

Jurgensen, R. C., Maier, J. E., & Donnelly, A. J. (1975). *Modern basic geometry* (Canadian metric ed.). Markham, ON, Canada: Houghton Mifflin.

Kelly, P., & Ladd, N. (1965). *Geometry.* Chicago: Scott Foresman and Co.

Kinsella, J. (1965). *Secondary school mathematics.* New York: The Center for Applied Research in Education.

Kline, M. (1961). *Why Johnny can't add: The failure of the new math.* New York: Random House.

Kline, M. (1953). *Mathematics in western culture.* Oxford, UK: Oxford University Press.

Klotz, E. (1991). Visualization in geometry: A case study of a multimedia mathematics education project. In W. Zimmerman & S. Cunningham (Eds.), *Visualization in teaching and learning mathematics* (pp. 95–104). Washington, DC: Mathematical Association of America.

Kortenkamp, U., & Richter-Gebert, J. (1999). *The interactive geometry software Cinderella.* Heidelberg: Springer-Verlag.

Laborde C. (1995). Designing tasks for learning geometry in a computer based environment. In L. Bruner & B. Jaworski (Eds.), *Technology in mathematics teaching: A bridge between teaching and learning,* (pp. 35–68). London: Chartwell-Bratt.

Lakoff, G., & Núñez, R. E. (2000). *Where mathematics comes from: How the embodied mind brings mathematics into being.* New York: Basic Books.

Lappan, G., Phillips, E., Fey, J., & Fitzgerald, W. (1996). *The Connected Mathematics Project.* http://www.math.msu.edu/cmp/

Lathrop, T., & Stevens, L. (1967). *Geometry: A contemporary approach.* Belmont, CA: Wadsworth.

Legendre, A. M. (1794). *Elements de géométrie.* Paris: Didot.

Legendre, A. M. (1823). *Elements de geometrie avec des notes* (12th ed.). Paris: Firmin Didot.

Lehrer, R., Jenkins, M., & Osana, H. (1998). Longitudinal study of children's reasoning about space and geometry. In R. Lehrer & D. Chazan (Eds.), *Designing learning environments for developing understanding of geometry and space* (pp. 137–167). Mahwah, NJ: Erlbaum.

Leisso, A., & Fisher, R. (1960). A survey of teachers' opinions of a revised mathematics curriculum. *Mathematics Teacher, 53*(2), 113–118.

Mallory, V., & Fehr, H. (1942). Mathematical education in wartime. *Mathematics Teacher, 35*(8), 291–298.

Mathematical Association. (1929). *The teaching of geometry in the schools.* London: G. Bell & Sons, Ltd.

Mayberry, J. (1983). The van Hiele levels of geometric thought in undergraduate preservice teachers. *Journal for Research in Mathematics Education, 14*(1), 58–69.

McCormick, C. (1929). *Teaching of general mathematics in secondary schools, contribution to education #386.* New York: Bureau of Publications, Teachers College, Columbia University.

McLeod, D. (1934). The use of locus problems in the teaching of geometry. *Mathematics Teacher, 27,* 336–339.

Meder, A. Jr. (1958). What is wrong with Euclid? *Mathematics Teacher, 51*(9), 578–84.

Meray, C. (1903). *Nouveaux éléments de géométrie*. Paris: Dijon. (Original work published 1874)

Moise, E. E. (1975). The meaning of Euclidean geometry in school mathematics. *Mathematics Teacher, 6*, 472-477.

Moise, E., & Downs, F. (1964). *Geometry*. Reading, MA: Addison-Wesley.

Moore, E. H. (1926). Foundations of mathematics. A general survey of progress in the last twenty-five years, *First Yearbook of the National Council of Teachers of Mathematics*. New York: Bureau of Publications, Teachers College, Columbia University.

Myers, G. W. (1910). *Second-year mathematics for secondary schools*. Chicago: University of Chicago Press.

National Advisory Committee on Mathematical Education (NACOME). (1975). *Overview and analysis of school mathematics, grades K-12*. Washington, DC: Conference Board of the Mathematical Sciences.

National Center for Education Statistics. (2003). *Digest of education statistics, 2002*. Web release, June 23. Elementary and secondary chapter. Retrieved September 22, 2004, from http://nces.ed.gov/pubs2003/2003060b.pdf.

National Committee on Mathematical Requirements. (1923). *The reorganization of mathematics in secondary education*. Boston: Houghton Mifflin.

National Council of Supervisors of Mathematics. (1978). Position statement on basic skills. *Mathematics Teacher, 71*(2) 147–152.

National Council of Teachers of Mathematics. (1930). *The teaching of geometry, NCTM yearbook*. Reston, VA: Author.

National Council of Teachers of Mathematics. (1973). *Geometry in the mathematics curriculum, NCTM yearbook*. Reston, VA: Author.

National Council of Teachers of Mathematics. (1980). *An agenda for action: Recommendations for school mathematics of the 1980s*. Reston, VA: Author.

National Council of Teachers of Mathematics. (1987). *Learning and teaching geometry K-12, NCTM yearbook*. Reston, VA: Author.

National Council of Teachers of Mathematics. (1989). *Curriculum and evaluation standards for school mathematics*. Reston, VA: Author.

National Council of Teachers of Mathematics. (1991). *Professional standards for teaching mathematics*. Reston, VA: Author.

National Council of Teachers of Mathematics. (2000). *Principles and standards for school mathematics*. Reston, VA: Author.

National Education Association. (1894). *Report of the committee of ten on secondary school studies*. New York: American Book Company.

National Education Association. (1912). *Final report of the national committee of fifteen on geometry syllabus*. Chicago: University of Chicago Press.

Netz, R. (1998). *The shaping of deduction in Greek mathematics: A study in cognitive history*. Cambridge, UK: Cambridge University Press.

Office of Education. (1966). *Digest of educational statistics—1966*. Washington, DC: U.S. Department of Health, Education, and Welfare.

Ostendorf, L. (1975). *Summary of course offering and enrollments in public secondary schools, 1972-73*. Washington, DC: U.S. Government Printing Office.

Papert, S. (1980). *Mindstorms: Children, computers, and powerful ideas*. New York: Basic Books.

Papy, G. (1964). *Groups*. New York: St. Martin's Press.

Pedoe, D. (1998). In love with geometry. *College Mathematics Journal, 29*(3), 170–188.

Perry, J. (1902). Teaching of mathematics. *Educational Review, 23*(2), 158–189.

Playfair, J. (1795). *Elements of geometry*. Edinburgh, UK: Nourse.

Pólya, G. (1954). *Mathematics and plausible reasoning: Induction and analogy in mathematics* (Vol. 1). Princeton, NJ: Princeton University Press.

Popkewitz, T. (2004). The alchemy of the mathematics curriculum: Inscription and the fabrication of the child. *American Educational Research Journal, 41*(1), 3–34.

Porter, A. (1989). A curriculum out of balance: The case of elementary school mathematics. *Educational Researcher, 18*(5), 9–15.

Quast, W. G. (1968). Geometry in the high schools of the United States: An historical analysis from 1890–1966. Unpublished doctoral dissertation, Rutgers, The State University of New Jersey.

Ravitch, D. (2000). *Left back: A century of failed school reforms*. New York: Simon & Schuster.

Reeve, W. (Ed.). (1930). *The teaching of geometry. Fifth Yearbook of the National Council of Teachers of Mathematics*. New York: Teachers College, Columbia University.

Reeve, W. (1936). Attacks on mathematics and how to meet them. In W. Reeve (Ed.), *The place of mathematics in modern education, Eleventh Yearbook of the National Council of Teachers of Mathematics* (pp. 1–21). New York: Teachers College, Columbia University.

Richards, J. (1988). *Mathematical visions: The pursuit of geometry in Victorian England*. Boston: Academic Press.

Russell, B. (1903). *The principles of mathematics*. Cambridge, UK: University Press.

Schnell, L., & Crawford, M. (1953). *Plane geometry: A clear thinking approach*. New York: McGraw-Hill.

School Mathematics Study Group. (1961b). *Mathematics for high school, Geometry*. New Haven, CT: Yale University Press.

School Mathematics Study Group. (1961a). *Geometry*. New Haven, CT: Yale University Press.

Schuster, S. (1967). Geometric transformations. In W. McNaff (Ed.), *Geometry in the secondary school* (pp. 29–38). Washington, DC: NCTM.

Schwartz, J., & Yerushalmy, M. (1985). *The Geometric Supposer series. Macintosh*. Pleasantville, NY: Sunburst Communications.

Senk, S. (1989). van Hiele levels and achievement in writing geometry proofs. *Journal for Research in Mathematics Education, 20*(3), 309–321.

Senk, S., & Thompson, D. (2003). *Standards-oriented school mathematics curricula: What does the research say about student outcomes?* Mahwah, NJ: Erlbaum.

Serra, M. (1989). *Discovering geometry: An inductive approach*. Berkeley, CA: Key Curriculum Press.

Shibli, J. (1932). *Recent developments in the teaching of geometry*. State College, Pennsylvania: J. Shibli.

Simson, R. (1781). *The elements of Euclid: viz. the first six books, together with the eleventh and twelfth*. Edinburgh, UK: J. Nourse.

Sinclair, N., & Crespo, S. (2006) Learning mathematics in dynamic computer environments. *Teaching Children Mathematics*, *12*(9), p. 436.

Sinclair, N., & Jackiw, N. (2005). Understanding and projecting ICT trends. In S. Johnston-Wilder & D. Pimm (Eds.), *Teaching secondary matheamtics effectively with technology* (pp. 235–252). Open University Press.

Sitomer, H. (1964). Coordinate geometry with an affine approach. *Mathematics Teacher*, *57*, 404–405.

Sizer, T. (1964). *Secondary schools at the turn of the century*. New Haven, CT: Yale University Press.

Smith, D. E. (1926). A general survey of the progress of mathematics in our high schools in the last twenty-five years. *First Yearbook of the National Council of Teachers of Mathematics*. New York: Teachers College, Columbia University.

Smith, D. E. (1928). Mathematics in the training for citizenship, *Selected topics in the teaching of mathematics. Third Yearbook of the National Council of Teachers of Mathematics*. New York: Teachers College, Columbia University.

Smith, D., & Ginsburg, J. (1934). *A history of mathematics in America before 1900*. Chicago: The Mathematics Association of America, with cooperation of the Open Court Publishing Company.

Spencer, W. (1876). *Inventional geometry*. Woodstock, GA: American Book Company.

Stamper, A. (1909). *A history of the teaching of elementary geometry, with reference to present-day problems*. Unpublished doctoral thesis, Columbia University.

Swenson, J. (1935). *Integrated mathematics with special applications to geometry*. Ann Arbor, MI: Edwards Brothers Inc.

Szabo, S. (1966). University of Illinois committee on school mathematics. *Science Education News*, AAAS Misc. Publ. No. 66-22(November), pp. 4–5.

Tahta, D. (1980). About geometry. *For the Learning of Mathematics*, *1*(1), 2–9.

Taylor, L. (1930). An introduction to demonstrative geometry. *Mathematics Teacher*, *23*(4), 227–235.

University of Chicago School Mathematics Project (UCSMP). (1990). *Algebra* (Teacher's ed.). Glenview, IL: Scott, Foresman.

University of Illinois Committee on School Mathematics. (1960). *High school mathematics. Unit 6, Geometry*. Urbana: University of Illinois Press.

Usiskin, Z. P. (1969). *Effects of teaching Euclidian geometry via transformations on student achievement and attitudes in tenth grade geometry*. Ann Arbor, MI: University of Michigan.

Usiskin, Z. P., & Griffin, J. (2007). *The classification of quadrilaterals: A study in definition*. Charlotte, NC: Information Age.

van Hiele, P. (1986). *Structure and insight: A theory of mathematics education*. Orlando, FL: Academic Press.

Welchons, A. M., & Krickenberger, W. R. (1933). *Plane geometry*. New York: Ginn & Co.

Welchons, A. M., & Krickenberger, W. R. (1956). *New plane geometry*. New York: Ginn & Co.

Wells, W., & Hart, W. W. (1935). *Progressive plane geometry*. Boston: D.C. Heath.

Wentworth, G. (1888). *A text-book of geometry*. Boston: Ginn & Company.

Wentworth, G. (1892). *Plane geometry*. Boston: Ginn and Company.

Wentworth, G., & Smith, D. (1910). *Wentworth's plane geometry.* Revised in 1913. New York: Ginn & Co.

Wentworth, G., & Smith, D. (1913). *Plane and solid geometry.* Boston: Ginn & Co.

Whiteley, W. (1999). The decline and rise of geometry in 20th century North America. In J. G. McLoughlin (Ed.), *Canadian mathematics study group conference proceedings* (p. 7–30). St John's: Memorial University of Newfoundland.

Willoughby, S. S. (1967). *Contemporary teaching of secondary school mathematics.* New York: Wiley.

Wooten, W. (1965). *SMSG The making of a curriculum.* New Haven, CT: Yale University Press.

Wren, F., & McDonough, H. (1934a). Development of mathematics in secondary schools of the United States. *Mathematics Teacher, 27*(6), 281–295.

Wren, F. & McDonough, H. (1934b). Development of mathematics in secondary schools of the United States. *Mathematics Teacher, 27*(3), 117–127.

Yerushalmy, M., Chazan, D., & Gordon, M. (1993). Posing problems: One aspect of bringing inquiry into classrooms. In J. L. Schwartz, M. Yerushalmy, & B. Wilson (Eds.), *The Geometric Supposer: What is it a case of?* (pp. 117–142). Hillsdale, NJ: Erlbaum.

Young, J. (1906). *The teaching of mathematics in the elementary and the secondary school.* New York: Longmans, Green & Co.

Young, J., & Schwartz, A. (1915). *Plane geometry.* New York: Holt.

ABOUT THE AUTHOR

Nathalie Sinclair is currently an assistant professor in the Faculty of Education at Simon Fraser University in Canada, and before that, worked in the Department of Mathematics at Michigan State University. Her interest in the history of geometry stems from her master's degree work on the influence of ancient Greek geometry on medieval Islamic mathematics.

Printed in the United States
143414LV00002B/2/P

9 781593 116965